HowExpert Guide to Cloud

Master the Art, Science, and Wonders of Skywatching for Cloud Identification

HowExpert

Copyright © 2024 Hot Methods, Inc. DBA HowExpert™
www.HowExpert.com

For more tips related to this topic, visit HowExpert.com/cloudspotting.

Recommended Resources

- HowExpert.com – How To Guides on All Topics from A to Z by Everyday Experts.
- HowExpert.com/free – Free HowExpert Email Newsletter.
- HowExpert.com/books – HowExpert Books
- HowExpert.com/courses – HowExpert Courses
- HowExpert.com/clothing – HowExpert Clothing
- HowExpert.com/membership – HowExpert Membership Site
- HowExpert.com/affiliates – HowExpert Affiliate Program
- HowExpert.com/jobs – HowExpert Jobs
- HowExpert.com/writers – Write About Your #1 Passion/Knowledge/Expertise & Become a HowExpert Author.
- HowExpert.com/resources – Additional HowExpert Recommended Resources
- YouTube.com/HowExpert – Subscribe to HowExpert YouTube.
- Instagram.com/HowExpert – Follow HowExpert on Instagram.
- Facebook.com/HowExpert – Follow HowExpert on Facebook.
- TikTok.com/@HowExpert – Follow HowExpert on TikTok.

Publisher's Foreword

Dear HowExpert Reader,

HowExpert publishes quick 'how to' guides on all topics from A to Z by everyday experts.

At HowExpert, our mission is to discover, empower, and maximize everyday people's talents to ultimately make a positive impact in the world for all topics from A to Z…one everyday expert at a time!

HowExpert guides are written by everyday people just like you and me, who have a passion, knowledge, and expertise for a specific topic.

We take great pride in selecting everyday experts who have a passion, real-life experience in a topic, and excellent writing skills to teach you about the topic you are also passionate about and eager to learn.

We hope you get a lot of value from our HowExpert guides, and it can make a positive impact on your life in some way. All of our readers, including you, help us continue living our mission of positively impacting the world for all spheres of influences from A to Z.

If you enjoyed one of our HowExpert guides, then please take a moment to send us your feedback from wherever you got this book.

Thank you, and I wish you all the best in all aspects of life.

To your success,

Byungjoon "BJ" Min 민병준
Founder & Publisher of HowExpert
HowExpert.com

PS…If you are also interested in becoming a HowExpert author, then please visit our website at HowExpert.com/writers. Thank you & again, all the best!
John 3:16

COPYRIGHT, LEGAL NOTICE AND DISCLAIMER:

COPYRIGHT © 2024 HOT METHODS, INC. (DBA HOWEXPERT™). ALL RIGHTS RESERVED WORLDWIDE. NO PART OF THIS PUBLICATION MAY BE REPRODUCED IN ANY FORM OR BY ANY MEANS, INCLUDING SCANNING, PHOTOCOPYING, OR OTHERWISE WITHOUT PRIOR WRITTEN PERMISSION OF THE COPYRIGHT HOLDER.

DISCLAIMER AND TERMS OF USE: PLEASE NOTE THAT MUCH OF THIS PUBLICATION IS BASED ON PERSONAL EXPERIENCE AND ANECDOTAL EVIDENCE. ALTHOUGH THE AUTHOR AND PUBLISHER HAVE MADE EVERY REASONABLE ATTEMPT TO ACHIEVE COMPLETE ACCURACY OF THE CONTENT IN THIS GUIDE, THEY ASSUME NO RESPONSIBILITY FOR ERRORS OR OMISSIONS. ALSO, YOU SHOULD USE THIS INFORMATION AS YOU SEE FIT, AND AT YOUR OWN RISK. YOUR PARTICULAR SITUATION MAY NOT BE EXACTLY SUITED TO THE EXAMPLES ILLUSTRATED HERE; IN FACT, IT'S LIKELY THAT THEY WON'T BE THE SAME, AND YOU SHOULD ADJUST YOUR USE OF THE INFORMATION AND RECOMMENDATIONS ACCORDINGLY.

THE AUTHOR AND PUBLISHER DO NOT WARRANT THE PERFORMANCE, EFFECTIVENESS OR APPLICABILITY OF ANY SITES LISTED OR LINKED TO IN THIS BOOK. ALL LINKS ARE FOR INFORMATION PURPOSES ONLY AND ARE NOT WARRANTED FOR CONTENT, ACCURACY OR ANY OTHER IMPLIED OR EXPLICIT PURPOSE.

ANY TRADEMARKS, SERVICE MARKS, PRODUCT NAMES OR NAMED FEATURES ARE ASSUMED TO BE THE PROPERTY OF THEIR RESPECTIVE OWNERS, AND ARE USED ONLY FOR REFERENCE. THERE IS NO IMPLIED ENDORSEMENT IF WE USE ONE OF THESE TERMS.

NO PART OF THIS BOOK MAY BE REPRODUCED, STORED IN A RETRIEVAL SYSTEM, OR TRANSMITTED BY ANY OTHER MEANS: ELECTRONIC, MECHANICAL, PHOTOCOPYING, RECORDING, OR OTHERWISE, WITHOUT THE PRIOR WRITTEN PERMISSION OF THE AUTHOR.

ANY VIOLATION BY STEALING THIS BOOK OR DOWNLOADING OR SHARING IT ILLEGALLY WILL BE PROSECUTED BY LAWYERS TO THE FULLEST EXTENT. THIS PUBLICATION IS PROTECTED UNDER THE US COPYRIGHT ACT OF 1976 AND ALL OTHER APPLICABLE INTERNATIONAL, FEDERAL, STATE AND LOCAL LAWS AND ALL RIGHTS ARE RESERVED, INCLUDING RESALE RIGHTS: YOU ARE NOT ALLOWED TO GIVE OR SELL THIS GUIDE TO ANYONE ELSE.

THIS PUBLICATION IS DESIGNED TO PROVIDE ACCURATE AND AUTHORITATIVE INFORMATION WITH REGARD TO THE SUBJECT MATTER COVERED. IT IS SOLD WITH THE UNDERSTANDING THAT THE AUTHORS AND PUBLISHERS ARE NOT ENGAGED IN RENDERING LEGAL, FINANCIAL, OR OTHER PROFESSIONAL ADVICE. LAWS AND PRACTICES OFTEN VARY FROM STATE TO STATE AND IF LEGAL OR OTHER EXPERT ASSISTANCE IS REQUIRED, THE SERVICES OF A PROFESSIONAL SHOULD BE SOUGHT. THE AUTHORS AND PUBLISHER SPECIFICALLY DISCLAIM ANY LIABILITY THAT IS INCURRED FROM THE USE OR APPLICATION OF THE CONTENTS OF THIS BOOK.

HOT METHODS, INC. DBA HOWEXPERT
EMAIL: SUPPORT@HOWEXPERT.COM
WEBSITE: WWW.HOWEXPERT.COM

**COPYRIGHT © 2024 HOT METHODS, INC. (DBA HOWEXPERT™)
ALL RIGHTS RESERVED WORLDWIDE.**

Table of Contents

Recommended Resources 2
Publisher's Foreword 3
Book Overview 15
Introduction to Cloud Spotting 21
   Welcome to the World of Cloud Spotting 21
      Why Cloud Spotting? 21
      What You Will Learn 21
      How to Use This Guide 22
      Final Thoughts 23
Chapter 1: Welcome to the World of Clouds 24
   1.1 The Joy of Cloud Spotting 24
      1.1.1 What is Cloud Spotting? 24
      1.1.2 The Benefits of Cloud Spotting 24
      1.1.3 A Brief History of Cloud Spotting 25
   1.2 How Cloud Spotting Enhances Your Life 27
      1.2.1 Relaxation and Mindfulness 27
      1.2.2 Educational Value 28
      1.2.3 Community and Social Connections 30
   1.3 Overview of the Book 31
      1.3.1 What You Will Learn 31
      1.3.2 How to Use This Book 33
      1.3.3 Additional Resources 34
   Chapter 1 Review 35
Part 1: The Magic of Clouds 39
Chapter 2: Why Clouds Matter 39
   2.1 The Role of Clouds in Weather and Climate 39
      2.1.1 Weather Patterns 39
      2.1.2 Climate Regulation 40
      2.1.3 The Water Cycle 41
      Practical Application 42
      Reflection and Next Steps 43
      Review Questions 43

2.2 Cultural and Historical Significance of Clouds 43
2.2.1 Myths and Legends 44
2.2.2 Clouds in Art and Literature 45
2.2.3 Historical Observations 46
Practical Application 47
Reflection and Next Steps 47
Review Questions 48
Conclusion 48
2.3 Personal Stories from Cloud Spotters 48
2.3.1 Inspiring Stories 48
2.3.2 Cloud Spotting Adventures 49
2.3.3 Community Contributions 50
Reflection and Next Steps 52
Review Questions 52
Conclusion 52
Chapter 2 Review 53
Chapter 3: Types of Clouds: An Introduction 56
3.1 High-Level Clouds 56
3.1.1 Cirrus 56
3.1.2 Cirrostratus 57
3.1.3 Cirrocumulus 58
Practical Application for High-Level Clouds 59
Reflection and Next Steps 60
Review Questions 61
Conclusion 61
3.2 Mid-Level Clouds 61
3.2.1 Altostratus 62
3.2.2 Altocumulus 63
Practical Application for Mid-Level Clouds 64
Reflection and Next Steps 65
Review Questions 65
Conclusion 66
3.3 Low-Level Clouds 66
3.3.1 Stratus 66

3.3.2 Stratocumulus 67

3.3.3 Nimbostratus 68

Practical Application for Low-Level Clouds 70

Reflection and Next Steps 71

Review Questions 71

Conclusion 71

3.4 Clouds with Vertical Development 71

3.4.1 Cumulus 72

3.4.2 Cumulonimbus 73

Practical Application for Clouds with Vertical Development 74

Reflection and Next Steps 75

Review Questions 76

Conclusion 76

Chapter 3 Review 76

Part 2: Mastering Cloud Identification 80

Chapter 4: Tools and Techniques for Cloud Spotting 80

4.1 Essential Gear for Cloud Spotting 80

4.1.1 Binoculars 80

4.1.2 Camera Equipment 81

4.1.3 Weather Apps 83

Practical Application for Cloud Spotting Gear 84

Reflection and Next Steps 85

Review Questions 85

Conclusion 86

4.2 Advanced Techniques for Cloud Spotting 86

4.2.1 Cloud Identification Apps 86

4.2.2 Photography Tips 87

4.2.3 Recording Observations 89

Practical Application for Advanced Cloud Spotting Techniques .. 90

Reflection and Next Steps 91

Review Questions 92

Conclusion 92

4.3 Creating a Cloud Spotting Journal 92

4.3.1 Setting Up Your Journal 92

- 4.3.2 What to Record 94
- 4.3.3 Tips for Consistent Entries 95
- Practical Application for Creating a Cloud Spotting Journal 96
- Reflection and Next Steps 98
- Review Questions 98
- Conclusion 98
- Chapter 4 Review 98

Chapter 5: In-Depth Cloud Identification 102
- 5.1 Recognizing Shapes, Textures, and Colors 102
  - 5.1.1 Identifying Features 102
  - 5.1.2 Color Changes 103
  - 5.1.3 Texture Variations 104
  - Practical Tips for Cloud Spotting 105
- 5.2 Seasonal and Regional Variations 106
  - 5.2.1 Clouds by Season 106
  - 5.2.2 Regional Cloud Patterns 108
  - 5.2.3 Special Weather Events 109
- 5.3 Understanding Cloud Dynamics and Formation 111
  - 5.3.1 Atmospheric Conditions 111
  - 5.3.2 Weather Fronts 112
  - 5.3.3 Air Pressure and Temperature 114
  - Practical Tips for Cloud Spotting 115
- Chapter 5 Review 116

Part 3: The Science of Clouds 120

Chapter 6: The Basics of Cloud Formation 120
- 6.1 The Water Cycle: Evaporation, Condensation, Precipitation 120
  - 6.1.1 Evaporation 120
  - 6.1.2 Condensation 121
  - 6.1.3 Precipitation 122
- 6.2 Atmospheric Layers and Their Role in Cloud Formation 124
  - 6.2.1 Troposphere 124
  - 6.2.2 Stratosphere 125
  - 6.2.3 Mesosphere 126
- 6.3 The Impact of Temperature, Pressure, and Humidity 127

    6.3.1 Temperature Variations 127

    6.3.2 Air Pressure Differences 128

    6.3.3 Humidity Levels 129

  Chapter 6 Review 131

Chapter 7: Advanced Meteorology for Cloud Spotters 134

  7.1 Meteorological Instruments and Their Uses 134

    7.1.1 Thermometers 134

    7.1.2 Barometers 136

    7.1.3 Anemometers 137

  7.2 Satellite Imagery and Doppler Radar 139

    7.2.1 Understanding Satellite Images 140

    7.2.2 Using Doppler Radar 142

    7.2.3 Interpreting Data 143

  7.3 Predicting Weather Through Clouds 145

    7.3.1 Cloud Patterns 145

    7.3.2 Weather Forecasting 147

    7.3.3 Practical Applications 149

  Chapter 7 Review 151

Part 4: Cloud Spotting as a Lifestyle 153

Chapter 8: Starting Your Cloud Spotting Journey 153

  8.1 Best Times and Places for Cloud Spotting 153

    8.1.1 Optimal Weather Conditions 153

    8.1.2 Ideal Locations 154

    8.1.3 Seasonal Considerations 155

  8.2 Cloud Spotting Alone vs. in Groups 156

    8.2.1 Benefits of Solo Cloud Spotting 156

    8.22 Advantages of Group Cloud Spotting 157

    8.2.3 Organizing Cloud Spotting Events 158

  8.3 Staying Safe While Cloud Spotting 159

    8.3.1 Weather Safety Tips 159

    8.3.2 Environmental Awareness 160

    8.3.3 Personal Safety Precautions 161

  Chapter 8 Review 162

Chapter 9: Capturing the Beauty of Clouds 166

- 9.1 Photography Tips and Tricks 166
  - 9.1.1 Camera Settings 166
  - 9.1.2 Composition Techniques 167
  - 9.1.3 Editing Tips 168
- 9.2 The Art of Sketching Clouds 169
  - 9.2.1 Materials Needed 169
  - 9.2.2 Basic Techniques 170
  - 9.2.3 Advanced Tips 171
- 9.3 Sharing Your Cloud Spotting Experiences Online 173
  - 9.3.1 Social Media Platforms 173
  - 9.3.2 Blogging Your Journey 174
  - 9.3.3 Building an Online Community 175
- Chapter 9 Review 177
  - 9.1 Photography Tips and Tricks 177
- Part 5: The Artistic and Cultural Side of Clouds 181
- Chapter 10: Clouds in Art and Media 181
  - 10.1 Famous Paintings and Photographs 181
    - 10.1.1 Iconic Paintings 181
    - 10.1.2 Renowned Photographers 183
    - 10.1.3 Artistic Movements 185
  - 10.2 Clouds in Literature and Film 187
    - 10.2.1 Literary Works 187
    - 10.2.2 Filmography 189
    - 10.2.3 Symbolism in Media 190
  - 10.3 How Clouds Inspire Creativity 192
    - 10.3.1 Artistic Expression 192
    - 10.3.2 Personal Creativity 193
    - 10.3.3 Collaborative Projects 193
  - Chapter 10 Review 194
- Chapter 11: Creating Cloud-Inspired Art 198
  - 11.1 DIY Cloud Art Projects 198
    - 11.1.1 Painting Clouds 198
    - 11.1.2 Sculpting Clouds 200
    - 11.1.3 Digital Cloud Art 201

11.2 Hosting a Cloud-Themed Art Show 203
    11.2.1 Planning and Organization 203
    11.2.2 Displaying Your Work 204
    11.2.3 Engaging Your Audience 206
11.3 Collaborating with Other Cloud Spotters 207
    11.3.1 Group Projects 208
    11.3.2 Community Art 209
    11.3.3 Online Collaborations 210
Chapter 11 Review 212
Part 6: Exploring Rare and Extraordinary Clouds 215
Chapter 12: Chasing Rare Cloud Formations 215
12.1 Identifying and Photographing Unique Clouds 215
    12.1.1 Rare Cloud Types 215
    12.1.2 Photography Tips 217
    12.1.3 Documentation Techniques 218
12.2 Stories from the Field: Extreme Cloud Spotting Adventures 219
    12.2.1 Personal Accounts 219
    12.2.2 Memorable Experiences 220
    12.2.3 Lessons Learned 222
12.3 Understanding Optical Phenomena: Halos, Sundogs, Rainbows 223
    12.3.1 Formation Processes 223
    12.3.2 Identification Tips 224
    12.3.3 Capturing Phenomena 225
Chapter 13: Cloud Spotting Around the World 228
13.1 Famous Cloud Spotting Locations Globally 228
    13.1.1 North America 228
    13.1.2 Europe 229
    13.1.3 Asia and Beyond 230
13.2 Travel Tips for Cloud Enthusiasts 232
    13.2.1 Planning Your Trip 232
    13.2.2 Essential Gear 233
    13.2.3 Local Considerations 234
13.3 Connecting with International Cloud Spotting Communities 236
    13.3.1 Global Networks 236

  13.3.2 Online Communities 237
  13.3.3 Collaborative Events 238
 Chapter 13 Review 239
Part 7: The Future of Cloud Spotting 241
Chapter 14: Clouds and Climate Change 241
 14.1 How Changing Climates Affect Clouds 241
  14.1.1 Impact on Cloud Formation 241
  14.1.2 Regional Variations 242
  14.1.3 Long-term Changes 243
 14.2 The Role of Citizen Science in Cloud Research 244
  14.2.1 Contribution Opportunities 244
  14.2.2 Data Collection 245
  14.2.3 Influencing Research 246
 14.3 Innovations in Cloud Observation 248
  14.3.1 New Technologies 248
  14.3.2 Emerging Trends 249
  14.3.3 Future Prospects 250
 Chapter 14 Review 251
Chapter 15: Inspiring Future Cloud Spotters 254
 15.1 Teaching Cloud Spotting to Kids 254
  15.1.1 Educational Activities 254
  15.1.2 Fun Projects 255
  15.1.3 Building Interest 256
  15.2.1 Starting a Club 258
  15.2.2 Organizing Events 259
  15.2.3 Engaging Members 260
 15.3 Building a Community of Cloud Enthusiasts 261
  15.3.1 Networking Tips 261
  15.3.2 Online Platforms 263
  15.3.3 Long-term Engagement 264
 Chapter 15 Review 265
Conclusion 268
Chapter 16: Embracing the Sky 268
 16.1 Reflecting on the Beauty and Importance of Clouds 268

  16.1.1 Personal Reflections ... 268
  16.1.2 The Joy of Observation ... 269
  16.1.3 Inspiring Others ... 270
 16.2 Encouraging Lifelong Cloud Spotting ... 271
  16.2.1 Maintaining Interest ... 272
  16.2.2 Continuous Learning ... 273
  16.2.3 Sharing Your Passion ... 274
 16.3 Final Thoughts and Inspiration ... 275
  16.3.1 Inspirational Quotes ... 275
  16.3.2 Encouraging Words ... 276
  16.3.3 Future of Cloud Spotting ... 277
 Chapter 16 Review ... 279
Appendices ... 281
Appendix A: Glossary of Cloud Terms ... 281
 A1 Comprehensive List of Cloud-Related Terms ... 281
 A2 Easy Reference for Beginners and Experts Alike ... 282
  Step 1: User-Friendly Format ... 282
  Step 2: Visual Aids ... 283
  Step 3: Practical Applications ... 283
  Step 4: Cross-References ... 283
Appendix B: Resources for Cloud Spotters ... 284
 B1 Recommended Books, Websites, and Apps ... 284
  Books ... 284
  Websites ... 284
  Apps ... 285
 B2 Cloud Spotting Organizations and Groups ... 285
  Cloud Appreciation Society ... 285
  American Meteorological Society (AMS) ... 285
  Royal Meteorological Society (RMetS) ... 286
  Local Meteorological Societies and Clubs ... 286
 B3 Further Reading and Learning Opportunities ... 286
  Online Courses and Webinars ... 286
  Publications and Journals ... 287
  Workshops and Conferences ... 287

  Citizen Science Projects 287
Appendix C: Cloud Spotting Checklist 288
 C1 Essential Items for Your Cloud Spotting Kit 288
  1. Binoculars 288
  2. Camera 288
  3. Notebook and Pen 288
  4. Cloud Identification Guide 289
  5. Weather Apps 289
  6. Sunscreen and Hat 289
  7. Comfortable Clothing and Footwear 289
  8. Water and Snacks 289
  9. Field Guide for Weather Phenomena 290
  10. Portable Chair or Blanket 290
 C2 Key Points for Successful Cloud Identification 290
  1. Learn the Basic Cloud Types 290
  2. Observe Cloud Shape and Structure 290
  3. Note the Cloud's Altitude 291
  4. Pay Attention to Color and Lighting 291
  5. Look for Weather Patterns 291
  6. Observe Movement and Change 291
  7. Use a Cloud Identification Guide 292
  8. Take Detailed Notes and Photos 292
  9. Compare with Known Examples 292
  10. Join Cloud Spotting Communities 292
 Appendix D: Personal Cloud Spotting Log Template 293
  D1 Customizable Templates for Recording Your Observations 293
  D2 Tips for Keeping an Organized and Detailed Log 295
About the Author 298
About the Publisher 299
Recommended Resources 300

Book Overview

Discover the enchanting world of clouds with the "HowExpert Guide to Cloud Spotting." This comprehensive guide covers everything you need to become an expert cloud spotter, from understanding cloud formations to capturing their beauty in art and photography.

Introduction to Cloud Spotting

- Welcome to the World of Cloud Spotting: Begin your journey into the fascinating practice of observing clouds.

- Why Cloud Spotting?: Learn the benefits and joys of cloud spotting.

- What You Will Learn: Overview of the skills and knowledge you'll gain.

- How to Use This Guide: Tips for navigating and making the most of this guide.

- Final Thoughts: Encouragement and inspiration for your cloud spotting journey.

Chapter 1: Welcome to the World of Clouds

- The Joy of Cloud Spotting: Discover what cloud spotting is and its benefits.

- How Cloud Spotting Enhances Your Life: Learn how it promotes relaxation, education, and social connections.

- Overview of the Book: Introduction to the book's structure and resources.

Part 1: The Magic of Clouds

- Chapter 2: Why Clouds Matter

  - Role in Weather and Climate: Understand how clouds influence weather patterns, climate regulation, and the water cycle.

- Cultural and Historical Significance: Explore clouds in myths, art, literature, and history.

- Personal Stories from Cloud Spotters: Read inspiring stories and adventures from cloud enthusiasts.

Chapter 3: Types of Clouds: An Introduction

- High-Level Clouds: Identify cirrus, cirrostratus, and cirrocumulus clouds.

- Mid-Level Clouds: Learn about altostratus and altocumulus clouds.

- Low-Level Clouds: Recognize stratus, stratocumulus, and nimbostratus clouds.

- Clouds with Vertical Development: Understand cumulus and cumulonimbus clouds.

Part 2: Mastering Cloud Identification

- Chapter 4: Tools and Techniques for Cloud Spotting

  - Essential Gear: Binoculars, camera equipment, weather apps, and more.

  - Advanced Techniques: Cloud identification apps, photography tips, and keeping a cloud spotting journal.

- Chapter 5: In-Depth Cloud Identification

  - Shapes, Textures, and Colors: Recognize features and variations in clouds.

  - Seasonal and Regional Variations: Understand how clouds differ by season and region.

  - Cloud Dynamics and Formation: Learn about atmospheric conditions, weather fronts, and air pressure.

Part 3: The Science of Clouds

- Chapter 6: The Basics of Cloud Formation

  - Water Cycle: Evaporation, condensation, and precipitation.

  - Atmospheric Layers: Troposphere, stratosphere, and mesosphere.

  - Impact of Temperature, Pressure, and Humidity: How these factors affect cloud formation.

- Chapter 7: Advanced Meteorology for Cloud Spotters

  - Meteorological Instruments: Thermometers, barometers, and anemometers.

  - Satellite Imagery and Doppler Radar: Using technology to predict weather.

  - Predicting Weather Through Clouds: Learn to forecast weather by observing cloud patterns.

Part 4: Cloud Spotting as a Lifestyle

- Chapter 8: Starting Your Cloud Spotting Journey

  - Best Times and Places: Optimal conditions and locations for cloud spotting.

  - Solo vs. Group Cloud Spotting: Benefits of both approaches and organizing events.

  - Safety Tips: Weather, environmental, and personal safety.

- Chapter 9: Capturing the Beauty of Clouds

  - Photography Tips: Camera settings, composition techniques, and editing.

- Sketching Clouds: Materials, basic techniques, and advanced tips.

  - Sharing Experiences Online: Social media, blogging, and building a community.

Part 5: The Artistic and Cultural Side of Clouds

- Chapter 10: Clouds in Art and Media

  - Famous Paintings and Photographs: Iconic artworks and photographers.

  - Literature and Film: How clouds are depicted in writing and movies.

  - Inspiring Creativity: Using clouds to spark artistic and personal creativity.

- Chapter 11: Creating Cloud-Inspired Art

  - DIY Projects: Painting, sculpting, and digital cloud art.

  - Hosting Art Shows: Planning, displaying, and engaging audiences.

  - Collaborating with Others: Group projects and community art.

Part 6: Exploring Rare and Extraordinary Clouds

- Chapter 12: Chasing Rare Cloud Formations

  - Identifying Unique Clouds: Tips for spotting and photographing rare formations.

  - Stories from the Field: Personal accounts and lessons from experienced cloud spotters.

  - Optical Phenomena: Understanding and capturing halos, sundogs, and rainbows.

- Chapter 13: Cloud Spotting Around the World

  - Famous Locations: Top spots for cloud spotting globally.

  - Travel Tips: Planning trips, essential gear, and local considerations.

  - Connecting with Communities: Global networks and online collaborations.

Part 7: The Future of Cloud Spotting

- Chapter 14: Clouds and Climate Change

  - Climate Impact: How changing climates affect clouds.

  - Citizen Science: Contributing to cloud research.

  - Innovations in Observation: New technologies and future trends.

- Chapter 15: Inspiring Future Cloud Spotters

  - Teaching Kids: Educational activities and projects.

  - Starting Clubs: Organizing and engaging members.

  - Building Communities: Networking and long-term engagement.

Conclusion: Embracing the Sky

- Reflecting on Clouds: The beauty and importance of cloud observation.

- Lifelong Cloud Spotting: Maintaining interest and sharing your passion.

- Final Thoughts: Inspirational quotes and future prospects.

Appendices

- Glossary of Cloud Terms: Easy reference for beginners and experts.

- Resources for Cloud Spotters: Recommended books, websites, apps, and organizations.

- Cloud Spotting Checklist: Essential items and tips for successful spotting.

- Personal Cloud Spotting Log Template: Templates and tips for organized observations.

Unlock the beauty and mystery of the skies with the "HowExpert Guide to Cloud Spotting." Whether you're a beginner or an experienced enthusiast, this guide provides all the tools and knowledge you need to elevate your cloud spotting adventures. Get your copy today and start exploring the ever-changing canvas above!

Introduction to Cloud Spotting

Welcome to the World of Cloud Spotting

Clouds have always been a source of fascination and inspiration for people across the ages. From the wispy cirrus clouds that seem to dance in the sky to the towering cumulonimbus clouds that signal impending storms, clouds are an ever-changing and dynamic feature of our atmosphere. In this guide, you will embark on a journey to become a proficient cloud spotter, learning not only how to identify different types of clouds but also how to appreciate the stories they tell about our weather and climate.

Why Cloud Spotting?

Cloud spotting is more than just a pastime; it's a way to connect with nature and understand the environment around us. By learning to recognize various cloud formations, you can:

1. Predict Weather: Certain clouds are indicators of specific weather conditions. Knowing what to look for can help you anticipate changes in the weather.

2. Enhance Photography: Clouds add drama and interest to photographs. Understanding cloud formations can improve your landscape photography skills.

3. Relax and Meditate: Observing clouds can be a calming and meditative activity, providing a moment of tranquility in our busy lives.

4. Educational Value: Cloud spotting is a fun and educational activity for all ages, fostering an interest in meteorology and the natural world.

What You Will Learn

In this guide, we will cover everything you need to know to become an expert cloud spotter. The content is organized systematically to ensure you gain a comprehensive understanding of clouds and their significance. Here's a glimpse of what you will find in this book:

1. The Basics of Cloud Formation: Learn about the science behind how clouds form and the factors that influence their shapes and sizes.

2. Types of Clouds: Detailed descriptions of the different cloud types, including their characteristics, formation processes, and what they indicate about the weather.

3. Cloud Identification Techniques: Step-by-step instructions on how to identify clouds, with tips on what to look for and how to use cloud charts.

4. Weather Prediction with Clouds: Understand how to use cloud observations to make short-term weather predictions.

5. Cloud Photography Tips: Practical advice on capturing the beauty of clouds through photography, including equipment recommendations and composition techniques.

6. Cloud Spotting Activities: Fun and engaging activities for individuals and groups to enhance your cloud spotting skills.

7. Resources and Tools: A comprehensive list of resources, including books, websites, and apps, to further your knowledge and enjoyment of cloud spotting.

How to Use This Guide

This guide is designed to be both informative and practical. Each chapter builds on the previous one, creating a structured learning path for you to follow. Whether you are a complete beginner or have some experience with cloud spotting, you will find valuable insights and techniques to deepen your understanding and appreciation of clouds.

Feel free to read the chapters in order or skip to the sections that interest you the most. Each chapter contains detailed explanations, practical tips, and illustrative photographs to help you grasp the concepts and apply them in real life.

__Final Thoughts__

Cloud spotting is a delightful and enriching activity that offers countless benefits. By taking the time to look up and observe the clouds, you open yourself to a world of beauty and wonder that is often overlooked. We hope this guide inspires you to explore the skies with curiosity and awe, and that you find joy and satisfaction in your cloud spotting adventures.

Let's embark on this journey together and unlock the secrets of the skies. Happy cloud spotting!

Chapter 1: Welcome to the World of Clouds

1.1 The Joy of Cloud Spotting

1.1.1 What is Cloud Spotting?

Cloud spotting, also known as cloud watching or cloud gazing, is the practice of observing and identifying different types of clouds in the sky. It's a fascinating hobby that blends elements of science, art, and mindfulness. Unlike traditional meteorology, which focuses on data collection and weather prediction, cloud spotting is about enjoying the beauty and diversity of clouds and understanding their role in the atmosphere.

To get started with cloud spotting, you don't need any specialized equipment. All you need is a comfortable place to sit or stand, a clear view of the sky, and a willingness to look up and appreciate the natural world. As you become more experienced, you might want to use binoculars or a camera to get a closer look at the intricate details of clouds.

Cloud spotting can be done anytime and anywhere, making it an accessible hobby for people of all ages. Whether you're in a bustling city, a quiet countryside, or near the ocean, clouds are always present, providing endless opportunities for observation and discovery.

1.1.2 The Benefits of Cloud Spotting

Cloud spotting offers a range of benefits, from relaxation and mindfulness to educational value and community building. Here's how cloud spotting can enhance your life:

A. Relaxation and Mindfulness

Cloud spotting is a peaceful and meditative activity that encourages you to slow down and focus on the present moment. Watching clouds drift across the sky can reduce stress, lower anxiety, and provide a sense of calm. It's an excellent way to practice mindfulness and connect with nature.

B. Educational Value

Observing clouds can teach you a lot about the weather and the atmosphere. By learning to identify different cloud types and understanding their formation processes, you gain insight into meteorology and climate science. This knowledge can help you make short-term weather predictions and deepen your appreciation of the natural world.

C. Community and Social Connections

Cloud spotting can be a social activity that brings people together. There are cloud spotting clubs and online communities where enthusiasts share their observations, photographs, and experiences. Joining these groups can provide a sense of belonging and an opportunity to connect with like-minded individuals.

D. Creativity and Inspiration

Clouds have inspired artists, poets, and dreamers for centuries. Their ever-changing shapes and colors can spark your creativity and provide inspiration for various forms of art, from photography and painting to writing and music. Cloud spotting encourages you to see the beauty in everyday moments and find artistic expression in nature.

1.1.3 A Brief History of Cloud Spotting

The history of cloud spotting is rich and varied, reflecting humanity's long-standing fascination with the skies.

A. Early Observations

Clouds have been observed and recorded for millennia. Ancient civilizations, including the Greeks and Romans, noted cloud formations and their implications for weather. The philosopher Aristotle, in his work "Meteorologica," described different types of clouds and their role in weather patterns.

B. The Birth of Modern Meteorology

In the 19th century, cloud classification became more systematic with the work of Luke Howard, an English meteorologist. In 1803, Howard introduced the first system for naming clouds, categorizing them into three main types: cirrus, cumulus, and stratus. This system laid the foundation for modern cloud classification and is still in use today, with additional subcategories and variations.

C. Cultural Significance

Throughout history, clouds have held significant cultural and artistic value. They have been depicted in paintings, literature, and music, symbolizing various emotions and themes. For example, the Romantic poets of the 18th and 19th centuries often used clouds to convey moods of melancholy, inspiration, and transcendence.

D. The Rise of Cloud Spotting as a Hobby

In the 20th century, cloud spotting emerged as a popular hobby. The formation of the Cloud Appreciation Society in 2005 by Gavin Pretor-Pinney brought together cloud enthusiasts from around the world, promoting the joy and educational value of cloud spotting. The society has since grown to include thousands of members, who share photographs, observations, and stories, fostering a global community of cloud watchers.

E. Technological Advances

Recent technological advancements have enhanced the cloud spotting experience. High-quality cameras, weather apps, and online platforms allow enthusiasts to capture, share, and discuss their observations more easily. Satellite imagery and weather radars also provide deeper insights into cloud dynamics and weather prediction.

Cloud spotting has evolved from early observations to a structured hobby with a global community. Understanding this history enriches the experience, providing context and depth to the practice of cloud spotting. As you continue to explore the skies, you join a long tradition of sky watchers who have found wonder and inspiration in the clouds above.

1.2 How Cloud Spotting Enhances Your Life

Cloud spotting is more than just a casual pastime; it offers numerous benefits that can significantly enhance your overall well-being. This section will explore how cloud spotting can contribute to relaxation and mindfulness, provide educational value, and foster community and social connections.

1.2.1 Relaxation and Mindfulness

Cloud spotting is a powerful tool for relaxation and mindfulness. Here's how you can use it to reduce stress and increase your sense of peace:

A. *Finding a Quiet Spot*

1. Choose Your Location: Find a quiet, comfortable place with an unobstructed view of the sky. Parks, beaches, or even your backyard can be ideal.

2. Settle In: Bring a blanket or chair to sit on. Make sure you are comfortable to fully immerse yourself in the experience.

B. Focusing on the Present Moment

1. Breathe Deeply: Start by taking a few deep breaths to center yourself. Pay attention to your breath as it enters and leaves your body.

2. Observe Without Judgment: Look up at the sky and observe the clouds. Notice their shapes, movements, and transformations without trying to label or categorize them initially.

C. Mindfulness Techniques

1. Describe the Clouds: Silently describe the clouds to yourself. Focus on their texture, color, and form.

2. Embrace the Moment: Allow yourself to be fully present. Let go of worries and thoughts about the past or future. Simply enjoy the beauty of the clouds.

D. Daily Practice

1. Set Aside Time: Incorporate cloud spotting into your daily routine, even if it's just for a few minutes.

2. Consistency: Regular practice can help develop a habit of mindfulness and relaxation.

1.2.2 Educational Value

Cloud spotting is not only relaxing but also an educational activity that enhances your understanding of the natural world. Here's how it can be a learning experience:

A. Understanding Weather Patterns

1. Learn Cloud Types: Start by familiarizing yourself with basic cloud types such as cumulus, cirrus, stratus, and nimbus. Use cloud charts and identification guides.

2. Weather Predictions: Different clouds can indicate specific weather conditions. For example, cirrus clouds might signal good weather, while cumulonimbus clouds often indicate thunderstorms.

B. Enhancing Observation Skills

1. Detailed Observations: Practice making detailed observations. Note the cloud's shape, altitude, and movement.

2. Record Keeping: Keep a cloud spotting journal where you record your observations, date, time, and weather conditions.

C. Engaging in Citizen Science

1. Participate in Projects: Join citizen science projects that involve cloud observations. Platforms like NASA's GLOBE Observer app allow you to contribute data to scientific research.

2. Learning from Experts: Engage with meteorologists and cloud enthusiasts through workshops, webinars, and online forums.

D. Integrating with Other Subjects

1. Cross-Disciplinary Learning: Use cloud spotting to enhance your knowledge in subjects like geography, environmental science, and art.

2. Educational Resources: Utilize books, documentaries, and online resources to deepen your understanding of meteorology and climatology.

1.2.3 Community and Social Connections

Cloud spotting can also help you connect with others who share your interest. Building a community around cloud spotting can enhance your experience and provide social benefits:

A. Joining Clubs and Organizations

1. Local Groups: Look for local cloud spotting clubs or nature groups. These organizations often organize outings and events.

2. Online Communities: Join online forums and social media groups dedicated to cloud spotting. These platforms are great for sharing photos, stories, and tips.

B. Participating in Events

1. Meetups and Workshops: Attend meetups, workshops, and conferences related to cloud spotting and meteorology.

2. Cloud Spotting Festivals: Participate in or organize local cloud spotting festivals to celebrate and share your passion with others.

C. Sharing Your Passion

1. Photography and Art: Share your cloud photographs and artwork with the community. Participate in exhibitions or online galleries.

2. Education and Outreach: Teach others about cloud spotting. Conduct sessions at schools, community centers, or through social media platforms.

D. Building Lifelong Friendships

1. Collaborative Projects: Work on collaborative projects with other cloud spotters, such as creating a cloud atlas or conducting joint observations.

2. Support Networks: Develop a support network of like-minded individuals who can share in your interests and experiences.

By integrating cloud spotting into your life, you can enjoy a more relaxed, informed, and socially connected lifestyle. Whether you are looking to unwind, learn more about the weather, or make new friends, cloud spotting offers a multitude of benefits that can enrich your daily experience.

1.3 Overview of the Book

In this section, we will provide a comprehensive overview of what you can expect from this guide. We'll outline the key learning objectives, explain how to navigate and use the book effectively, and list additional resources to enhance your cloud spotting journey.

1.3.1 What You Will Learn

This book is designed to transform you into an adept cloud spotter, equipped with the knowledge and skills to appreciate and understand the beauty and science of clouds. Here's what you will learn:

A. The Basics of Cloud Formation

- Understanding Cloud Types: You will learn to identify various cloud types, such as cirrus, cumulus, stratus, and more.

- Cloud Formation Processes: Gain insight into how clouds form, including the roles of temperature, humidity, and atmospheric pressure.

- Weather Patterns and Predictions: Discover how different cloud types can help you predict weather changes.

B. Practical Cloud Spotting Techniques

- Observation Skills: Develop your ability to observe and note cloud characteristics accurately.

- Using Tools and Technology: Learn how to use tools like binoculars, cameras, and weather apps to enhance your cloud spotting experience.

- Recording Observations: Keep detailed records of your observations in a cloud spotting journal.

C. Cloud Spotting as a Lifestyle

- Mindfulness and Relaxation: Use cloud spotting as a tool for relaxation and mindfulness.

- Community Engagement: Connect with fellow cloud enthusiasts through clubs, events, and online communities.

- Creative Expression: Explore ways to express your cloud spotting experiences through photography, art, and writing.

D. Advanced Meteorological Concepts

- In-Depth Cloud Identification: Delve deeper into identifying subtle differences between cloud types.

- Meteorological Instruments: Learn about the tools meteorologists use to study clouds and weather.

- Climate Change Impact: Understand how climate change affects cloud formation and behavior.

1.3.2 How to Use This Book

To maximize your learning and enjoyment, it's essential to know how to navigate and use this book effectively. Here's a structured approach:

A. Structured Learning Path

- Follow the Chapters Sequentially: Each chapter builds on the previous one, providing a structured learning path. Start from the basics and gradually move to more advanced topics.

- Use Chapter Summaries: At the end of each chapter, summaries are provided to reinforce key points and concepts.

B. Practical Application

- Exercises and Activities: Engage in the practical exercises and activities included in each chapter. These are designed to help you apply what you've learned in real-world cloud spotting.

- Observation Logs: Utilize the templates provided for maintaining a cloud spotting journal. Recording your observations systematically will enhance your skills and understanding.

C. Reference Sections

- Glossary of Terms: Refer to the glossary for definitions of key terms and concepts related to cloud spotting and meteorology.

- Resource Lists: Check the resource sections for recommended books, websites, apps, and organizations that can provide additional information and support.

D. Interactive Elements

- Online Communities: Join online forums and social media groups mentioned in the book to share your experiences and learn from others.

- Field Guide Integration: Use the book as a field guide during your cloud spotting outings. The detailed descriptions and images will help you identify and understand different cloud types on the go.

1.3.3 Additional Resources

To further enhance your cloud spotting journey, we have compiled a list of additional resources. These will provide you with more in-depth knowledge and tools to expand your understanding and enjoyment of clouds:

A. Books and Field Guides

- "The Cloudspotter's Guide" by Gavin Pretor-Pinney: A comprehensive and entertaining guide to cloud spotting.

- "The Cloud Book" by Tomie dePaola: A beautifully illustrated book that explains various cloud types.

B. Websites and Online Platforms

- Cloud Appreciation Society (cloudappreciationsociety.org): An excellent resource for cloud enthusiasts, offering a wealth of information and a community of cloud spotters.

- NASA's GLOBE Observer App (observer.globe.gov): Participate in citizen science projects by submitting your cloud observations to NASA.

C. Apps and Technology

- Cloud Identification Apps: Apps like "CloudSpotter" can help you identify clouds and learn more about them.

- Weather Apps: Apps such as "Weather Underground" provide real-time weather data and forecasts, which can be useful for cloud spotting.

D. Community and Organizations

- Local Cloud Spotting Clubs: Join local clubs or nature groups that focus on cloud spotting and meteorology.

- Online Forums and Social Media Groups: Engage with other cloud enthusiasts on platforms like Reddit, Facebook, and Instagram.

E. Educational Workshops and Courses

- Meteorology Workshops: Attend workshops or online courses to deepen your understanding of weather and clouds.

- Photography Classes: Take classes to improve your cloud photography skills, capturing the beauty of the sky more effectively.

This overview provides a roadmap for your journey through the world of cloud spotting. By following the structured approach outlined in this book, you will develop a thorough understanding and appreciation of clouds, enhancing both your knowledge and enjoyment. Happy cloud spotting!

Chapter 1 Review

1.1 The Joy of Cloud Spotting

- What is Cloud Spotting?: The practice of observing and identifying different types of clouds, focusing on their beauty and atmospheric role. It requires no specialized equipment and can be done anywhere with a clear view of the sky.

- The Benefits of Cloud Spotting: Provides relaxation and mindfulness, enhances educational understanding of weather, fosters community and social connections, and inspires creativity.

- A Brief History of Cloud Spotting: From early observations by ancient civilizations to the development of modern meteorology by Luke Howard, cultural significance in art and literature, the rise of cloud spotting as a hobby, and the impact of technological advancements.

1.2 How Cloud Spotting Enhances Your Life

- Relaxation and Mindfulness: Find a quiet spot, practice mindfulness by observing clouds without judgment, and incorporate cloud spotting into your daily routine.

- Educational Value: Learn to identify cloud types and predict weather changes, enhance observation skills, keep a cloud spotting journal, and participate in citizen science projects.

- Community and Social Connections: Join local and online cloud spotting communities, attend and organize events and workshops, and share your passion through photography, art, and teaching others.

1.3 Overview of the Book

- What You Will Learn: Basics of cloud formation and identification, practical cloud spotting techniques, integrating cloud spotting into your lifestyle, and advanced meteorological concepts.

- How to Use This Book: Follow chapters sequentially, engage in practical exercises, use observation logs, utilize glossary and resource lists, and join online communities.

- Additional Resources: Recommended books, websites, apps, cloud spotting organizations, forums, and educational workshops and classes.

Practical Application

1. Prepare Your Equipment: Gather binoculars, a camera, and a notebook.

2. Find Your Spot: Choose a location with a clear view of the sky.

3. Start Observing: Look up and note different cloud types.

4. Record Your Observations: Keep a detailed journal of your findings.

5. Join a Community: Engage with local and online cloud spotting groups.

Reflection and Next Steps

Reflect on what you've learned in this chapter, set goals for your cloud spotting journey, and prepare for the next chapter, which will delve deeper into the types of clouds and their significance.

Review Questions

1. What is cloud spotting and why is it beneficial?

2. How can cloud spotting contribute to relaxation and mindfulness?

3. What educational benefits does cloud spotting offer?

4. How can joining a community enhance your cloud spotting experience?

5. What key elements will you learn from this book?

Conclusion

This chapter has provided an introduction to cloud spotting, its benefits, and an overview of what you will learn. With this foundation, you are now ready to explore the art and science of cloud spotting in greater depth in the upcoming chapters. Happy cloud spotting!

Part 1: The Magic of Clouds

Chapter 2: Why Clouds Matter

Clouds are not just picturesque elements of our sky; they play crucial roles in weather and climate systems. Understanding these roles is essential for appreciating the significance of clouds in our environment.

2.1 The Role of Clouds in Weather and Climate

2.1.1 Weather Patterns

Clouds are integral to the weather patterns we experience daily. Here's how they influence weather:

A. Types of Weather-Influencing Clouds

- Cumulus Clouds: These fluffy, white clouds typically indicate fair weather, but when they grow larger and darker, they can develop into cumulonimbus clouds, bringing thunderstorms.

- Stratus Clouds: These low, gray clouds cover the sky like a blanket and are often associated with overcast weather and light rain or drizzle.

- Cirrus Clouds: High, wispy clouds that usually signal a change in the weather, such as an approaching warm front.

B. Cloud Movement and Wind Patterns

- Wind Direction: The movement of clouds can indicate wind direction and speed at various altitudes.

- Weather Fronts: The type and movement of clouds can signify the approach of weather fronts, which are boundaries between different air masses. For example, a line of cumulonimbus clouds often marks a cold front.

C. Precipitation Indicators

- Rain Clouds: Nimbostratus and cumulonimbus clouds are typically associated with precipitation. Nimbostratus clouds bring continuous, steady rain, while cumulonimbus clouds can produce heavy rain, hail, and thunderstorms.

- Snow Clouds: In colder climates, stratus and nimbostratus clouds can lead to snowfall.

2.1.2 Climate Regulation

Clouds play a significant role in regulating the Earth's climate by affecting the energy balance:

A. Reflecting Solar Radiation

- Albedo Effect: Clouds reflect a portion of the incoming solar radiation back into space. This cooling effect, known as the albedo effect, helps moderate the Earth's temperature.

- Cloud Types and Albedo: Different cloud types have varying albedo effects. For example, thick, low-lying clouds like stratus clouds reflect more sunlight compared to high, thin clouds like cirrus clouds.

B. Trapping Heat

- Greenhouse Effect: Clouds also trap heat radiating from the Earth's surface, contributing to the greenhouse effect. This warming effect helps maintain the planet's temperature, particularly during the night.

- Cloud Coverage and Temperature: Areas with significant cloud cover tend to have less temperature variation between day and night, as clouds prevent rapid cooling after sunset.

C. Feedback Mechanisms

- Climate Feedback Loops: Clouds are involved in complex feedback loops that can either amplify or dampen climate changes. For instance, increased cloud cover due to higher temperatures can lead to more reflection of solar radiation, potentially offsetting some warming.

- Modeling Challenges: Predicting cloud behavior is a major challenge in climate modeling due to the complexity of cloud formation and their interactions with other atmospheric elements.

2.1.3 The Water Cycle

Clouds are a vital component of the Earth's water cycle, which is crucial for sustaining life:

A. Evaporation

- Water Vapor Formation: Water from oceans, lakes, rivers, and soil evaporates into the atmosphere, forming water vapor. This process is driven by solar energy.

- Humidity and Cloud Formation: High levels of humidity in the air contribute to cloud formation when water vapor condenses into tiny droplets or ice crystals.

B. Condensation

- Cloud Formation: As moist air rises and cools, the water vapor condenses onto particles like dust, salt, or smoke, forming clouds. This process releases latent heat, which can influence weather patterns.

- Types of Clouds Formed: The altitude and temperature at which condensation occurs determine the type of clouds formed, such as cumulus, stratus, or cirrus clouds.

C. Precipitation

- Rain, Snow, Sleet, and Hail: When cloud droplets combine and grow large enough, they fall to the Earth as precipitation. The type of precipitation depends on the temperature profile of the atmosphere.

- Distribution of Freshwater: Precipitation is essential for distributing freshwater across the planet, replenishing lakes, rivers, and aquifers.

D. Collection

- Surface Runoff: Precipitation that falls on land flows into rivers, lakes, and oceans, contributing to surface runoff. Some of it infiltrates the soil, recharging groundwater supplies.

- Cycle Continuation: This collected water eventually evaporates again, continuing the water cycle.

Practical Application

Understanding the role of clouds in weather and climate is crucial for cloud spotting. Here are some practical steps to apply this knowledge:

1. Observe and Identify: When you see clouds, try to identify their type and consider what weather they might indicate.

2. Track Weather Changes: Keep a log of cloud observations and correlate them with subsequent weather changes to improve your predictive skills.

3. Use Technology: Utilize weather apps and satellite images to compare your observations with professional data.

4. Engage with the Community: Share your observations with cloud spotting communities to learn from others and refine your understanding.

Reflection and Next Steps

Reflect on how clouds influence weather and climate. Consider how this knowledge enhances your cloud spotting skills. Prepare for the next chapter, where you will delve deeper into the types of clouds and their characteristics.

Review Questions

1. How do clouds influence weather patterns?

2. What roles do clouds play in climate regulation?

3. How are clouds involved in the water cycle?

4. What are some practical applications of understanding cloud roles in weather and climate?

Conclusion

This chapter has provided an in-depth look at why clouds matter, emphasizing their roles in weather, climate regulation, and the water cycle. With this knowledge, you are now better equipped to appreciate the significance of clouds and apply this understanding to your cloud spotting endeavors. Happy cloud spotting!

2.2 Cultural and Historical Significance of Clouds

Clouds have fascinated humans for centuries, serving as symbols in myths and legends, subjects in art and literature, and objects of scientific observation. This section explores the rich cultural and historical significance of clouds.

2.2.1 Myths and Legends

Clouds have been woven into the fabric of myths and legends across various cultures, often symbolizing divine presence, weather phenomena, and moral lessons. Here are some notable examples:

A. Greek and Roman Mythology

- Zeus and the Clouds: In Greek mythology, clouds were often associated with Zeus, the god of the sky and thunder. He was believed to control the weather, sending clouds to bring rain or storms.

- Nephology: The study of clouds, known as nephology, derives its name from "Nephele," a cloud nymph in Greek mythology. Nephele was created by Zeus from a cloud to test the faithfulness of Ixion.

B. Norse Mythology

- The Giants' Breath: In Norse mythology, clouds were thought to be the breath of giants, reflecting their close connection to the natural world and the elements.

- Yggdrasil and Weather: The great tree Yggdrasil was believed to hold up the sky, with its branches influencing weather patterns, including the formation and movement of clouds.

C. Native American Legends

- Rain Spirits: Many Native American tribes revered clouds as manifestations of rain spirits. For instance, the Hopi tribe believed that kachina spirits lived in the clouds and brought rain to nourish the earth.

- Sky Beings: Some tribes, such as the Lakota, viewed clouds as the dwelling place of sky beings who controlled the weather and were integral to their spiritual practices.

2.2.2 Clouds in Art and Literature

Clouds have been a profound source of inspiration in art and literature, symbolizing emotions, themes, and natural beauty. Here are some notable contributions:

A. Renaissance and Baroque Art

- Dramatic Skies: Artists like Leonardo da Vinci and Michelangelo depicted clouds with dramatic flair, using them to enhance the emotional intensity of their paintings. Clouds were often used to signify divine presence or intervention.

- Baroque Grandeur: Baroque artists such as Peter Paul Rubens and Caravaggio used clouds to create dynamic compositions, adding movement and depth to their works.

B. Romanticism

- Nature's Majesty: Romantic artists like J.M.W. Turner and Caspar David Friedrich focused on the sublime aspects of nature, including the grandeur of clouds. Their works often portrayed clouds as vast, powerful forces that dwarfed human endeavors.

- Emotional Expression: Clouds in Romantic literature, such as the poetry of William Wordsworth and Percy Bysshe Shelley, were used to express emotions ranging from awe to melancholy.

C. Modern and Contemporary Art

- Impressionism: Artists like Claude Monet captured the fleeting beauty of clouds, emphasizing light and color. His series of paintings, including "Haystacks" and "Water Lilies," often feature intricate cloudscapes.

- Abstract and Conceptual Art: Contemporary artists like Anish Kapoor and Olafur Eliasson use clouds and cloud-like forms in abstract and conceptual installations, exploring themes of perception and environmental change.

D. Literature

- Symbolism in Poetry: Clouds have been a recurring motif in poetry, symbolizing change, transience, and introspection. Poets like John Keats and Emily Dickinson used clouds to evoke mood and metaphor.

- Narrative Elements: In novels and stories, clouds often set the scene or reflect the internal states of characters. For example, in Gabriel García Márquez's "One Hundred Years of Solitude," clouds are used to symbolize the passage of time and the cyclical nature of history.

2.2.3 Historical Observations

The study of clouds has a rich history, with significant contributions from early observers to modern meteorologists. Here are some key milestones:

A. Ancient Observations

- Aristotle's Meteorologica: Aristotle's work "Meteorologica," written in 350 BCE, was one of the earliest attempts to systematically describe weather phenomena, including clouds. He classified clouds based on their appearance and altitude.

- Chinese and Indian Scholars: Ancient Chinese and Indian scholars also made significant contributions to the understanding of clouds and weather patterns, often linking their observations to agricultural practices.

B. Luke Howard's Classification

- Founding Father of Cloud Classification: In 1803, Luke Howard, an English pharmacist and amateur meteorologist, introduced the first systematic classification of clouds. His nomenclature, which included terms like cirrus, cumulus, and stratus, is still used today.

- Scientific Recognition: Howard's work gained recognition among contemporary scientists and was instrumental in advancing the study of meteorology.

C. Advancements in Meteorology

- The 20th Century: The advent of photography and later, satellite imagery, revolutionized the study of clouds. These technologies allowed for more precise observation and analysis of cloud formations and their impact on weather and climate.

- Modern Research: Today, meteorologists use advanced tools such as Doppler radar and satellite data to study clouds. These technologies provide detailed information about cloud composition, movement, and their role in atmospheric processes.

Practical Application

Understanding the cultural and historical significance of clouds can enhance your appreciation and enjoyment of cloud spotting. Here are some practical steps to integrate this knowledge into your cloud spotting practice:

1. Research and Reflect: Read about the myths, legends, and artistic depictions of clouds to deepen your understanding and connection to this natural phenomenon.

2. Visit Art Exhibits: Explore museums and galleries that feature artworks depicting clouds. Take note of how different artists interpret and represent clouds.

3. Historical Context: Study the contributions of early observers like Aristotle and Luke Howard to appreciate the evolution of cloud science.

4. Share Insights: Discuss your findings with fellow cloud spotters and engage in conversations about the cultural and historical aspects of clouds.

Reflection and Next Steps

Reflect on how clouds have been perceived and represented throughout history. Consider how this understanding can enrich your cloud spotting experience. Prepare for the next chapter, where we will delve into the types of clouds and their characteristics.

Review Questions

1. How have clouds been represented in myths and legends?

2. What role do clouds play in art and literature?

3. What are some key historical milestones in the observation and study of clouds?

4. How can understanding the cultural and historical significance of clouds enhance your cloud spotting practice?

Conclusion

This section has explored the rich cultural and historical significance of clouds, highlighting their impact on mythology, art, literature, and scientific observation. By understanding these perspectives, you can deepen your appreciation of clouds and enhance your cloud spotting journey. Happy cloud spotting!

2.3 Personal Stories from Cloud Spotters

Cloud spotting has not only captivated everyday enthusiasts but also played significant roles in the lives of some renowned historical figures. Their stories illustrate how clouds have influenced scientific discoveries, artistic creations, and historical events. Here are some famous factual accounts that highlight the impact of cloud spotting.

2.3.1 Inspiring Stories

A. Luke Howard: The Father of Cloud Classification

In the early 19th century, Luke Howard, an English pharmacist and amateur meteorologist, revolutionized the way we understand clouds. His pioneering work led to the first systematic classification of clouds, which he published in 1803. Howard's system categorized clouds into types such as cirrus, cumulus, and stratus, laying the foundation for modern meteorology. His passion for observing and understanding clouds not only advanced scientific knowledge but also inspired poets like Goethe, who admired Howard's work and wrote poems about clouds. Howard's story is a testament to how dedicated observation and classification can lead to significant scientific contributions.

B. John Constable: Capturing the Sky in Art

John Constable, a famous English landscape painter of the 19th century, is renowned for his detailed and expressive depictions of clouds. Constable believed that the sky was the "chief organ of sentiment" in a landscape, and his works often featured dramatic cloudscapes that captured the varying moods and weather conditions. He meticulously studied clouds and kept a cloud diary, recording his observations with sketches and notes. Constable's dedication to cloud spotting helped him achieve a level of realism and emotional depth in his paintings that continues to be celebrated today. His story highlights the artistic inspiration that clouds can provide.

C. Alfred Wegener: Clouds and Continental Drift

Alfred Wegener, a German polar researcher, geophysicist, and meteorologist, is best known for his theory of continental drift. During his expeditions to Greenland in the early 20th century, Wegener meticulously observed the clouds and weather patterns. These observations were crucial for understanding the climatic conditions of the polar regions. Wegener's interdisciplinary approach, combining meteorology with geology, helped him develop groundbreaking ideas about the Earth's structure. His story demonstrates how cloud observation can contribute to broader scientific theories and discoveries.

2.3.2 Cloud Spotting Adventures

A. William Rankin: The Man Who Fell Through a Thunderstorm

In 1959, William Rankin, a U.S. Marine Corps pilot, survived a harrowing fall through a thunderstorm after ejecting from his aircraft at high altitude. During his fall, Rankin experienced extreme weather conditions, including freezing temperatures, violent updrafts, and hailstones. His descent through the storm clouds provided unique firsthand insights into the behavior of cumulonimbus clouds and the dynamics of thunderstorms. Rankin's adventure, detailed in his book "The Man Who Rode the Thunder," offers a dramatic account of the power and complexity of storm clouds.

B. The Montgolfier Brothers: Pioneers of Ballooning

Joseph-Michel and Jacques-Étienne Montgolfier, French brothers who were aviation pioneers, conducted the first manned hot air balloon flight in 1783. Their flights often took them through various cloud layers, giving them a unique perspective on cloud formations and weather conditions at different altitudes. The Montgolfier brothers' experiments with ballooning not only advanced human flight but also contributed to early meteorological observations. Their story illustrates the adventurous spirit of exploring the skies and the importance of clouds in the development of aviation.

C. Felix Baumgartner: Jumping from the Edge of Space

In 2012, Felix Baumgartner, an Austrian skydiver, set a world record by jumping from the stratosphere, nearly 39 kilometers (24 miles) above the Earth. During his descent, Baumgartner passed through various atmospheric layers, including thin, high-altitude clouds. His jump provided valuable data on the behavior of the human body during supersonic freefall and contributed to our understanding of the upper atmosphere. Baumgartner's story underscores the intersection of extreme sports, scientific research, and cloud observation.

2.3.3 Community Contributions

A. NASA and Cloud Observation

NASA has long relied on citizen scientists to help gather cloud data. Programs like the GLOBE Observer app enable people worldwide to contribute to climate research by submitting their cloud observations. These contributions help validate satellite data and improve weather models. The collective effort of thousands of cloud spotters provides critical information that enhances our understanding of the Earth's climate. This story highlights the power of community contributions to scientific research.

B. The Cloud Appreciation Society

Founded in 2005 by Gavin Pretor-Pinney, the Cloud Appreciation Society has grown into a global community of cloud enthusiasts. The society celebrates the beauty and diversity of clouds, encouraging members to share photographs, observations, and personal reflections. The society's efforts have raised awareness about the importance of clouds in our ecosystem and inspired a greater appreciation for the sky. The Cloud Appreciation Society exemplifies how a community of passionate individuals can make a significant impact on public awareness and appreciation of natural phenomena.

C. The International Cloud Atlas

The International Cloud Atlas, first published in 1896 by the World Meteorological Organization (WMO), is a comprehensive reference for cloud classification and observation. It was created with contributions from meteorologists around the world, compiling a standardized system for identifying and describing clouds. The atlas has been an invaluable tool for weather professionals and enthusiasts alike, providing a common language for discussing cloud formations. This collective effort underscores the importance of international collaboration in advancing our understanding of clouds.

Practical Application

To enhance your cloud spotting experience, consider the following steps inspired by these historical stories:

1. Study Cloud Classification: Like Luke Howard, take the time to learn and apply cloud classification systems to improve your identification skills.

2. Keep a Cloud Diary: Follow John Constable's example by maintaining a journal of your cloud observations, including sketches and notes.

3. Engage in Citizen Science: Participate in programs like NASA's GLOBE Observer app to contribute to scientific research.

4. Explore Cloud Art and Literature: Draw inspiration from historical figures by exploring how clouds have been depicted in art and literature.

Reflection and Next Steps

Reflect on how these historical stories resonate with your own experiences and interests. Consider how you can incorporate elements of scientific inquiry, artistic inspiration, and community engagement into your cloud spotting practice. Prepare for the next chapter, where you will learn about the different types of clouds and their characteristics.

Review Questions

1. How did Luke Howard contribute to the classification of clouds?

2. What role did clouds play in John Constable's art?

3. How can citizen science projects enhance our understanding of clouds?

4. How have historical figures used cloud observation in their scientific and artistic endeavors?

Conclusion

This section has highlighted the personal stories of famous historical figures who have significantly contributed to the study and appreciation of clouds. By learning from these narratives, you can find inspiration and practical ways to enhance your own cloud spotting journey. Happy cloud spotting!

Chapter 2 Review

2.1 The Role of Clouds in Weather and Climate

- Weather Patterns: Clouds indicate and influence weather. Cumulus clouds suggest fair weather, while cumulonimbus clouds indicate storms. Observing clouds helps predict wind patterns and weather changes.

- Climate Regulation: Clouds reflect solar radiation (cooling effect) and trap heat (warming effect), maintaining Earth's temperature balance.

- The Water Cycle: Clouds form from water vapor, condense, and precipitate as rain, snow, etc., continuing the water cycle.

2.2 Cultural and Historical Significance of Clouds

- Myths and Legends: Clouds feature in myths from Greek, Roman, Norse, and Native American cultures, symbolizing gods, giants, and spirits.

- Clouds in Art and Literature: Artists like Constable and Monet, and writers like Wordsworth, used clouds to convey emotions and themes. Clouds remain a source of artistic inspiration.

- Historical Observations: Aristotle's early observations, Luke Howard's cloud classification, and advancements in meteorology through photography and satellite imagery.

2.3 Personal Stories from Cloud Spotters

- Inspiring Stories: Luke Howard classified clouds, John Constable painted them, and Alfred Wegener combined meteorology with geology.

- Cloud Spotting Adventures: William Rankin's fall through a thunderstorm, the Montgolfier Brothers' balloon flights, and Felix Baumgartner's jump from the stratosphere.

- Community Contributions: NASA's GLOBE Observer app, the Cloud Appreciation Society, and the International Cloud Atlas.

Practical Application

1. Study Cloud Classification: Learn cloud types to improve identification skills.

2. Keep a Cloud Diary: Record observations and sketches.

3. Engage in Citizen Science: Participate in programs like NASA's GLOBE Observer.

4. Explore Cloud Art and Literature: Draw inspiration from artistic and literary depictions of clouds.

Reflection and Next Steps

Reflect on the roles of clouds in weather, climate, and culture. Consider how this knowledge enhances your cloud spotting skills. Prepare for the next chapter on types of clouds and their characteristics.

Review Questions

1. How do clouds influence weather patterns and climate regulation?

2. What roles do clouds play in myths, legends, art, and literature?

3. Who were key historical figures in cloud observation, and what were their contributions?

4. How can personal and community contributions enhance cloud spotting?

Conclusion

This chapter covered the significance of clouds in weather, climate, culture, and personal stories of famous cloud spotters. Use this knowledge to deepen your appreciation of clouds and enhance your cloud spotting journey. Happy cloud spotting!

Chapter 3: Types of Clouds: An Introduction

Understanding the different types of clouds is fundamental to cloud spotting. This chapter introduces the various cloud types, starting with high-level clouds, which are typically found at altitudes above 20,000 feet (6,000 meters). High-level clouds are composed primarily of ice crystals due to the cold temperatures at such heights. Let's explore these clouds in detail.

3.1 High-Level Clouds

High-level clouds, also known as cirriform clouds, are typically wispy and delicate in appearance. They form at high altitudes and are indicators of weather changes. There are three main types of high-level clouds: cirrus, cirrostratus, and cirrocumulus.

3.1.1 Cirrus

Description:

- Cirrus clouds are thin, wispy, and often appear as delicate strands or patches in the sky.

- They are usually white and are composed of ice crystals.

Formation:

- Form at altitudes above 20,000 feet (6,000 meters) where the air is cold and dry.

- Created by the lifting of warm, moist air into the upper atmosphere.

Weather Indication:

- Typically indicate fair weather but can also signal that a change in the weather is coming, especially if they thicken or lower.

Identification Tips:

1. Look for Wispy Shapes: Cirrus clouds often look like wisps of hair or feathers.

2. Notice the High Altitude: They appear high in the sky and do not cast shadows on the ground.

3. Observe Movement: They move quickly due to strong upper-level winds.

Practical Application:

- Use cirrus clouds as an early indicator of approaching weather systems, such as a warm front.

3.1.2 Cirrostratus

Description:

- Cirrostratus clouds form a thin, milky veil across the sky, often covering it entirely or in large patches.

- These clouds are so thin that they usually allow sunlight or moonlight to pass through, creating halos around the sun or moon.

Formation:

- Develop from the gradual lifting of a large air mass, causing widespread ice crystal formation at high altitudes.

Weather Indication:

- Often precede a warm front and may indicate that precipitation is coming within the next 24 hours.

Identification Tips:

1. Look for a Uniform Layer: Cirrostratus clouds appear as a consistent, thin layer across the sky.

2. Notice the Halo Effect: The presence of a halo around the sun or moon is a clear sign of cirrostratus clouds.

3. Check for Shadows: They are high enough to not cast noticeable shadows on the ground.

Practical Application:

- Use the appearance of cirrostratus clouds as a signal to prepare for changing weather, especially precipitation.

3.1.3 Cirrocumulus

Description:

- Cirrocumulus clouds consist of small, white patches that appear in rows or ripples, often resembling a fish scale pattern known as a "mackerel sky."

- These clouds are less common than cirrus or cirrostratus.

Formation:

- Form at high altitudes from the turbulent mixing of air layers, causing the ice crystals to group together in small clusters.

Weather Indication:

- Usually indicate fair, but cold weather. Their presence can also suggest moisture and instability at high altitudes.

Identification Tips:

1. Look for Small Patches: Cirrocumulus clouds appear as small, white patches or ripples in the sky.

2. Check for Pattern: The mackerel sky pattern is a distinctive feature.

3. High Altitude: These clouds do not cast shadows and appear at high altitudes.

Practical Application:

- Use cirrocumulus clouds to anticipate cooler temperatures and possible changes in weather patterns.

Practical Application for High-Level Clouds

To effectively use your knowledge of high-level clouds in cloud spotting, follow these steps:

1. Observation:

- Regularly observe the sky and note the presence and types of high-level clouds.

- Use binoculars for a closer look at the patterns and textures of these clouds.

2. Recording:

- Keep a cloud diary to log your observations, including the time, date, and weather conditions.

- Sketch the clouds or take photographs to help with later identification and comparison.

3. Weather Prediction:

- Use high-level clouds as indicators of upcoming weather changes.

- Note any transitions from one cloud type to another, as this can signal shifts in weather patterns.

4. Community Engagement:

- Share your observations with cloud spotting communities online or in local groups.

- Compare your findings with weather forecasts to improve your predictive skills.

Reflection and Next Steps

Reflect on the characteristics and significance of high-level clouds. Consider how this knowledge can enhance your cloud spotting experience and weather prediction abilities. Prepare for the next section, where we will explore mid-level clouds and their features.

Review Questions

1. What are the primary characteristics of cirrus clouds?

2. How can cirrostratus clouds signal an approaching weather change?

3. What unique pattern do cirrocumulus clouds form, and what does it indicate about the weather?

4. How can you use high-level clouds to predict upcoming weather changes?

Conclusion

High-level clouds play a crucial role in weather prediction and understanding atmospheric conditions. By learning to identify cirrus, cirrostratus, and cirrocumulus clouds, you can enhance your cloud spotting skills and gain valuable insights into the weather. Continue to observe, record, and share your findings to deepen your appreciation and knowledge of the sky. Happy cloud spotting!

3.2 Mid-Level Clouds

Mid-level clouds typically form between 6,500 and 20,000 feet (2,000 to 6,000 meters) above the Earth's surface. They are composed of water droplets but can also contain ice crystals at higher altitudes or in colder climates. Understanding mid-level clouds is essential for accurate weather prediction and cloud spotting. This section will detail the characteristics, formation, and significance of altostratus and altocumulus clouds.

3.2.1 Altostratus

Description:

- Altostratus clouds are gray or blue-gray clouds that usually cover the entire sky, often making the sun or moon appear as if seen through frosted glass.

- These clouds are thicker and darker than high-level cirrostratus clouds.

Formation:

- Form from the gradual lifting of a large air mass that causes widespread condensation.

- Often associated with warm fronts and can precede precipitation.

Weather Indication:

- Altostratus clouds typically indicate that a storm with continuous rain or snow might be approaching within the next 12 to 24 hours.

- If they thicken and lower, they may transition into nimbostratus clouds, which bring steady precipitation.

Identification Tips:

1. Sky Coverage: Look for clouds that cover most or all of the sky.

2. Sun Appearance: The sun or moon may appear as a dim, watery disk through these clouds.

3. Uniform Color: These clouds often have a uniform gray or blue-gray color without distinct features or patterns.

Practical Application:

- Use altostratus clouds as indicators of approaching significant weather changes, particularly continuous precipitation.

3.2.2 Altocumulus

Description:

- Altocumulus clouds appear as white or gray patches that cover the sky in large, rounded masses or rolls. These masses often have shading and may look like cotton balls.

- They can appear as a field of patches or a single layer and are typically less dense and lower than cirrocumulus clouds.

Formation:

- Form from the turbulent mixing of air at mid-level altitudes.

- Can develop in various weather conditions, often signaling moisture and instability in the middle atmosphere.

Weather Indication:

- The presence of altocumulus clouds can indicate fair weather but can also signal a potential change within the next day, especially if they appear in the morning.

- Sometimes associated with thunderstorms later in the day if the air at higher levels is moist and unstable.

Identification Tips:

1. Patchy Appearance: Look for clouds that appear in patches or rows with gaps between them.

2. Shading and Texture: Notice the shaded and textured appearance, indicating variations in thickness.

3. Morning Presence: Pay special attention to altocumulus clouds in the morning, as they can indicate weather changes later in the day.

Practical Application:

- Use altocumulus clouds to anticipate potential weather changes, particularly the development of thunderstorms or shifts in weather patterns.

Practical Application for Mid-Level Clouds

To effectively use your knowledge of mid-level clouds in cloud spotting, follow these steps:

1. Observation:

- Regularly observe the sky and note the presence and types of mid-level clouds.

- Use binoculars for a closer look at the patterns and textures of these clouds.

2. Recording:

- Keep a cloud diary to log your observations, including the time, date, and weather conditions.

- Sketch the clouds or take photographs to help with later identification and comparison.

3. Weather Prediction:

- Use mid-level clouds as indicators of upcoming weather changes.

- Note any transitions from one cloud type to another, as this can signal shifts in weather patterns.

4. Community Engagement:

- Share your observations with cloud spotting communities online or in local groups.

- Compare your findings with weather forecasts to improve your predictive skills.

Reflection and Next Steps

Reflect on the characteristics and significance of mid-level clouds. Consider how this knowledge can enhance your cloud spotting experience and weather prediction abilities. Prepare for the next section, where we will explore low-level clouds and their features.

Review Questions

1. What are the primary characteristics of altostratus clouds?

2. How can altocumulus clouds signal potential weather changes?

3. What unique patterns do altocumulus clouds form, and what do they indicate about the weather?

4. How can you use mid-level clouds to predict upcoming weather changes?

Conclusion

Mid-level clouds play a crucial role in weather prediction and understanding atmospheric conditions. By learning to identify altostratus and altocumulus clouds, you can enhance your cloud spotting skills and gain valuable insights into the weather. Continue to observe, record, and share your findings to deepen your appreciation and knowledge of the sky. Happy cloud spotting!

3.3 Low-Level Clouds

Low-level clouds typically form below 6,500 feet (2,000 meters) and are primarily composed of water droplets, although they can contain ice particles during cold weather. Understanding low-level clouds is crucial for predicting short-term weather changes and precipitation. This section will detail the characteristics, formation, and significance of stratus, stratocumulus, and nimbostratus clouds.

3.3.1 Stratus

Description:

- Stratus clouds form a uniform gray layer that covers the sky like a blanket.

- These clouds often result in overcast conditions and can create a foggy or misty appearance at the ground level.

Formation:

- Form when a large air mass cools at the surface, causing widespread condensation at low altitudes.

- Often associated with gentle, steady lifting of moist air.

Weather Indication:

- Stratus clouds usually indicate overcast skies and can produce light mist or drizzle.

- They often signal stable weather conditions but can persist for extended periods.

Identification Tips:

1. Uniform Layer: Look for a featureless, gray sky that lacks distinct shapes or patterns.

2. Low Altitude: These clouds are close to the ground and may create fog-like conditions.

3. Consistent Coverage: Stratus clouds typically cover the entire sky uniformly.

Practical Application:

- Use stratus clouds to anticipate prolonged overcast conditions with light precipitation.

3.3.2 Stratocumulus

Description:

- Stratocumulus clouds appear as low, lumpy, gray or white clouds, often in patches or rolls with blue sky visible between them.

- They can vary in thickness and may have darker areas indicating more moisture.

Formation:

- Form from the mixing of cooler air with warmer, moist air at low altitudes.

- Often develop after a cold front has passed, as the air stabilizes.

Weather Indication:

- Typically indicate fair weather with possible periods of light rain or drizzle.

- Can signal improving weather conditions if they break up and disperse.

Identification Tips:

1. Patchy Appearance: Look for clouds that appear in large, lumpy patches or rolls.

2. Variable Color: Notice the varying shades of gray and white, indicating differences in cloud thickness.

3. Gaps of Blue Sky: Look for areas of blue sky visible between the cloud patches.

Practical Application:

- Use stratocumulus clouds to anticipate fair weather with occasional light precipitation, often indicating stable conditions.

3.3.3 Nimbostratus

Description:

- Nimbostratus clouds are thick, dark gray clouds that cover the entire sky and often bring continuous, steady precipitation.

- They are dense and lack distinct shapes, creating a dull, overcast sky.

Formation:

- Form from the extensive lifting of moist air, often associated with warm fronts.

- These clouds develop when a warm, moist air mass rises over a colder air mass, leading to widespread condensation and precipitation.

Weather Indication:

- Nimbostratus clouds are strong indicators of continuous, steady precipitation, such as rain or snow, lasting for extended periods.

- They often signify prolonged periods of dull, overcast weather.

Identification Tips:

1. Thick, Dark Layer: Look for a uniformly dark gray sky with no distinct cloud shapes.

2. Continuous Precipitation: These clouds bring steady rain or snow without heavy downpours.

3. Low Visibility: Nimbostratus clouds often reduce visibility due to their thickness and the associated precipitation.

Practical Application:

- Use nimbostratus clouds to anticipate extended periods of steady precipitation, often associated with warm fronts and prolonged overcast conditions.

Practical Application for Low-Level Clouds

To effectively use your knowledge of low-level clouds in cloud spotting, follow these steps:

1. Observation:

- Regularly observe the sky and note the presence and types of low-level clouds.

- Use binoculars to examine the textures and formations of these clouds more closely.

2. Recording:

- Keep a cloud diary to log your observations, including the time, date, and weather conditions.

- Sketch the clouds or take photographs to help with later identification and comparison.

3. Weather Prediction:

- Use low-level clouds to predict short-term weather changes and precipitation.

- Note any transitions from one cloud type to another, as this can signal shifts in weather patterns.

4. Community Engagement:

- Share your observations with cloud spotting communities online or in local groups.

- Compare your findings with weather forecasts to improve your predictive skills.

Reflection and Next Steps

Reflect on the characteristics and significance of low-level clouds. Consider how this knowledge can enhance your cloud spotting experience and weather prediction abilities. Prepare for the next chapter, where we will explore clouds with vertical development and their features.

Review Questions

1. What are the primary characteristics of stratus clouds?

2. How can stratocumulus clouds indicate fair weather or light precipitation?

3. What weather conditions are associated with nimbostratus clouds?

4. How can you use low-level clouds to predict upcoming weather changes?

Conclusion

Low-level clouds play a crucial role in weather prediction and understanding atmospheric conditions. By learning to identify stratus, stratocumulus, and nimbostratus clouds, you can enhance your cloud spotting skills and gain valuable insights into short-term weather changes. Continue to observe, record, and share your findings to deepen your appreciation and knowledge of the sky. Happy cloud spotting!

3.4 Clouds with Vertical Development

Clouds with vertical development are some of the most visually striking and dynamically significant cloud types. These clouds can form at various altitudes and grow vertically, often indicating active weather systems and significant weather changes. This section details the characteristics, formation, and significance of cumulus and cumulonimbus clouds.

3.4.1 Cumulus

Description:

- Cumulus clouds are fluffy, white clouds with a clearly defined edge, often resembling cotton balls or heaps of whipped cream.

- These clouds typically have a flat base and can appear isolated or in groups.

Formation:

- Form due to the convection of warm, moist air rising from the surface. As the air rises, it cools and condenses to form clouds.

- Often develop on sunny days when the ground heats up and causes air to rise.

Weather Indication:

- Cumulus clouds usually indicate fair weather but can grow into larger clouds that signal changing weather.

- When cumulus clouds grow vertically and become taller, they can develop into cumulonimbus clouds, indicating potential storms.

Identification Tips:

1. Fluffy Appearance: Look for clouds with a puffy, white appearance and a well-defined edge.

2. **Flat Base:** Cumulus clouds have a distinct flat base, often due to the uniform level at which condensation occurs.

3. **Vertical Growth:** Observe if the cloud is growing vertically, which might indicate developing weather changes.

Practical Application:

- Use cumulus clouds to predict fair weather but be aware of their potential to grow into cumulonimbus clouds, which can bring storms.

3.4.2 Cumulonimbus

Description:

- Cumulonimbus clouds are towering, dense clouds that can reach great heights, often appearing as massive, anvil-shaped formations.

- These clouds are dark and menacing at the base, with a brilliant white top due to the ice crystals at higher altitudes.

Formation:

- Form from rapidly rising warm, moist air, which cools and condenses as it ascends, leading to the development of a towering cloud structure.

- Typically associated with strong convection currents, often ahead of or during a cold front.

Weather Indication:

- Cumulonimbus clouds indicate severe weather, including thunderstorms, heavy rain, hail, and even tornadoes.

- The presence of these clouds signals highly unstable atmospheric conditions.

Identification Tips:

1. Towering Structure: Look for tall, towering clouds with a flat, dark base and a wide, anvil-shaped top.

2. Dark Base: The base of cumulonimbus clouds is usually dark due to the density of water droplets and ice particles.

3. Thunderstorm Activity: Observe for signs of lightning, heavy rain, or hail, which are common with cumulonimbus clouds.

Practical Application:

- Use cumulonimbus clouds as a clear indicator of severe weather. Take precautions if you see these clouds forming, as they can produce dangerous weather conditions.

Practical Application for Clouds with Vertical Development

To effectively use your knowledge of clouds with vertical development in cloud spotting, follow these steps:

1. Observation:

- Regularly observe the sky and note the presence and types of vertically developing clouds.

- Use binoculars for a closer look at the structure and details of these clouds.

2. Recording:

- Keep a cloud diary to log your observations, including the time, date, and weather conditions.

- Sketch the clouds or take photographs to help with later identification and comparison.

3. Weather Prediction:

- Use vertically developing clouds to predict significant weather changes and severe weather events.

- Note any transitions from cumulus to cumulonimbus clouds, as this signals increasing atmospheric instability.

4. Safety Precautions:

- Be aware of the potential for severe weather with cumulonimbus clouds and take necessary precautions, such as seeking shelter during thunderstorms.

5. Community Engagement:

- Share your observations with cloud spotting communities online or in local groups.

- Compare your findings with weather forecasts to improve your predictive skills and stay informed about potential severe weather.

Reflection and Next Steps

Reflect on the characteristics and significance of clouds with vertical development. Consider how this knowledge can enhance your cloud spotting experience and improve your ability to predict severe weather. Prepare for the next chapter, where we will delve into more advanced cloud spotting techniques and tools.

Review Questions

1. What are the primary characteristics of cumulus clouds?

2. How can cumulus clouds indicate potential weather changes?

3. What severe weather conditions are associated with cumulonimbus clouds?

4. How can you use vertically developing clouds to predict upcoming weather changes?

Conclusion

Clouds with vertical development play a crucial role in weather prediction and understanding atmospheric conditions. By learning to identify cumulus and cumulonimbus clouds, you can enhance your cloud spotting skills and gain valuable insights into significant weather changes. Continue to observe, record, and share your findings to deepen your appreciation and knowledge of the sky. Happy cloud spotting!

Chapter 3 Review

3.1 High-Level Clouds

- Cirrus: Wispy, thin clouds at high altitudes indicating fair weather or an approaching change. Look for delicate strands high in the sky.

- Cirrostratus: Thin, milky clouds covering the sky, creating halos around the sun or moon, often indicating a coming weather change.

- Cirrocumulus: Small, white patches in rows or ripples, resembling fish scales, often signaling fair but cool weather.

3.2 Mid-Level Clouds

- Altostratus: Gray or blue-gray clouds covering the sky, making the sun or moon appear as if through frosted glass, often indicating approaching precipitation.

- Altocumulus: White or gray patches or rolls, with blue sky visible between them, often indicating fair weather but can signal changes if seen in the morning.

3.3 Low-Level Clouds

- Stratus: Uniform gray clouds covering the sky like a blanket, leading to overcast conditions and light mist or drizzle.

- Stratocumulus: Low, lumpy clouds in patches or rolls with gaps of blue sky, typically indicating fair weather with possible light rain.

- Nimbostratus: Thick, dark gray clouds covering the sky, bringing continuous, steady precipitation, often associated with prolonged overcast conditions.

3.4 Clouds with Vertical Development

- Cumulus: Fluffy, white clouds with a flat base, indicating fair weather but can grow into larger clouds signaling changing weather.

- Cumulonimbus: Towering, dense clouds with an anvil-shaped top, indicating severe weather such as thunderstorms, heavy rain, hail, and tornadoes.

Practical Application

1. Observation:

   - Regularly observe the sky and note the presence of different cloud types across all altitudes.

- Use binoculars for detailed observation of cloud structures and formations.

2. Recording:

- Keep a detailed cloud diary to log observations, including time, date, and weather conditions.

- Sketch or photograph clouds to assist with identification and pattern recognition.

3. Weather Prediction:

- Use knowledge of cloud types to predict weather changes, noting transitions and developments in cloud formations.

- Be particularly mindful of vertically developing clouds for severe weather alerts.

4. Safety Precautions:

- Stay informed about weather conditions associated with cumulonimbus clouds, and take precautions during severe weather events.

5. Community Engagement:

- Share your observations with cloud spotting communities, both online and locally.

- Compare your findings with weather forecasts to refine your predictive skills and stay informed.

Reflection and Next Steps

Reflect on the characteristics and significance of each cloud type covered in this chapter. Consider how this knowledge enhances your cloud spotting experience and your ability to predict weather changes. Prepare for the next chapter, where advanced cloud spotting techniques and tools will be explored.

Review Questions

1. What are the distinguishing features of cirrus, cirrostratus, and cirrocumulus clouds?

2. How do altostratus and altocumulus clouds signal approaching weather changes?

3. What weather conditions are typically associated with stratus, stratocumulus, and nimbostratus clouds?

4. How can you use cumulus and cumulonimbus clouds to predict significant weather changes?

Conclusion

This chapter has provided a comprehensive overview of various cloud types across different altitudes, highlighting their characteristics, formation, and significance. By understanding these clouds, you can enhance your cloud spotting skills and gain valuable insights into weather prediction. Continue to observe, record, and share your findings to deepen your appreciation and knowledge of the sky. Happy cloud spotting!

Part 2: Mastering Cloud Identification

Chapter 4: Tools and Techniques for Cloud Spotting

Mastering cloud identification involves more than just knowing the different types of clouds. It requires the right tools and techniques to observe, document, and understand the clouds effectively. This chapter will guide you through the essential gear needed for cloud spotting and how to use it effectively.

4.1 Essential Gear for Cloud Spotting

Equipping yourself with the right tools can significantly enhance your cloud spotting experience. Here's a detailed look at the essential gear for cloud spotting:

4.1.1 Binoculars

Purpose:

- Binoculars allow you to observe clouds in greater detail, making it easier to identify specific characteristics and features that may not be visible to the naked eye.

A. Choosing Binoculars

1. Magnification: Opt for binoculars with a magnification of 7x to 10x. This range provides a good balance between detail and stability.

2. Field of View: A wider field of view is beneficial for scanning large portions of the sky. Look for binoculars with a field of view of at least 300 feet at 1000 yards.

3. Lens Quality: High-quality lenses with anti-reflective coatings provide clearer and brighter images. Multi-coated lenses are ideal.

B. Using Binoculars

1. Adjust the Focus: Start by focusing on a distant object on the horizon, then move to the clouds. Use the central focus wheel to get a sharp image.

2. Stabilize Your View: Hold the binoculars steady by resting your elbows on a solid surface or using a tripod if necessary.

3. Scan the Sky: Move slowly across the sky, pausing to examine interesting cloud formations in detail.

C. Practical Tips

1. Keep Clean: Keep your binoculars clean and dry. Use a lens cloth to remove dust and fingerprints.

2. Practice: Practice using your binoculars to become familiar with their operation and focusing mechanism.

4.1.2 Camera Equipment

Purpose:

- A camera allows you to capture and document your cloud observations. Photographs can be used for later analysis, comparison, and sharing with the cloud spotting community.

A. Choosing Camera Equipment

1. Type of Camera: A DSLR or mirrorless camera with manual settings is ideal for capturing high-quality images. However, a good smartphone camera can also be effective.

2. Lens: A wide-angle lens (16-35mm) is useful for capturing expansive sky scenes, while a telephoto lens (70-200mm) can help you focus on specific cloud details.

3. Tripod: A sturdy tripod is essential for stabilizing your camera, especially in low light conditions or when using a telephoto lens.

B. Using Camera Equipment

1. Camera Settings: Use a low ISO setting (100-200) to reduce noise and ensure sharp images. Adjust the aperture (f/8 to f/16) to get a good depth of field. Set a fast shutter speed (1/250 or faster) to capture moving clouds.

2. Composition: Frame your shots to include points of interest such as trees, buildings, or the horizon. This provides scale and context to your cloud photographs.

3. Lighting: Clouds look different under various lighting conditions. Experiment with photographing clouds at different times of the day, including sunrise, sunset, and during different weather conditions.

C. Practical Tips

1. Review and Edit: Regularly review and edit your photos to improve your technique and understand the characteristics of different cloud types.

2. Protect Gear: Keep your camera gear clean and protected from the elements. Use a weather-sealed camera bag.

4.1.3 Weather Apps

Purpose:

- Weather apps provide real-time data and forecasts, helping you plan your cloud spotting activities. They can also offer detailed information about cloud formations, weather patterns, and atmospheric conditions.

A. Choosing Weather Apps

1. AccuWeather: Provides detailed weather forecasts, including cloud cover predictions and satellite imagery.

2. Weather Underground: Offers hyper-local weather forecasts, radar maps, and cloud cover information.

3. CloudSpotter: Specifically designed for cloud enthusiasts, this app helps identify cloud types and track observations.

B. Using Weather Apps

1. Check Forecasts: Use weather apps to check the cloud cover and weather forecasts for your location. This helps you plan the best times for cloud spotting.

2. Satellite Imagery: View satellite images to see cloud formations over a wide area. This can help you understand larger weather patterns and predict cloud movement.

3. Radar Maps: Use radar maps to track precipitation and storm developments, particularly useful for spotting cumulonimbus clouds and other storm-related formations.

C. Practical Tips

1. Regular Updates: Regularly update your weather apps to ensure you have the latest data and features.

2. Cross-Reference: Use multiple weather apps to cross-reference information and get a more accurate picture of current and upcoming weather conditions.

Practical Application for Cloud Spotting Gear

To effectively use your cloud spotting gear, follow these steps:

1. Preparation:

- Gather your binoculars, camera equipment, and weather apps before heading out for cloud spotting.

- Ensure your gear is in good working condition and fully charged (if applicable).

2. Observation:

- Use your binoculars to scan the sky and identify cloud types in detail.

- Capture photographs of interesting cloud formations, adjusting your camera settings as needed.

3. Documentation:

- Keep a cloud spotting journal to record your observations, including the date, time, weather conditions, and cloud types observed.

- Use weather apps to annotate your observations with relevant data and forecasts.

4. Analysis:

- Review your photographs and notes to analyze cloud formations and weather patterns.

- Compare your findings with weather forecasts and satellite imagery to enhance your understanding of cloud behavior.

5. Sharing:

- Share your observations and photographs with cloud spotting communities online or in local groups.

- Engage with other enthusiasts to learn from their experiences and improve your cloud spotting skills.

Reflection and Next Steps

Reflect on how the right tools and techniques can enhance your cloud spotting experience. Consider investing in high-quality gear and using weather apps to plan your activities more effectively. Prepare for the next chapter, where we will delve into advanced techniques for identifying and understanding different cloud types.

Review Questions

1. What are the key features to look for when choosing binoculars for cloud spotting?

2. How can camera equipment enhance your cloud spotting experience?

3. What are the benefits of using weather apps in cloud spotting?

4. How can you effectively use your cloud spotting gear to observe, document, and analyze clouds?

Conclusion

Having the right tools and techniques is essential for mastering cloud spotting. By equipping yourself with binoculars, camera equipment, and weather apps, you can observe and document clouds in greater detail, enhancing your understanding and appreciation of the sky. Continue to practice and refine your skills, and share your findings with the cloud spotting community. Happy cloud spotting!

4.2 Advanced Techniques for Cloud Spotting

Building on the foundational gear, advanced techniques can greatly enhance your cloud spotting experience. This section covers the use of cloud identification apps, photography tips for capturing clouds, and methods for recording your observations systematically.

4.2.1 Cloud Identification Apps

Purpose:

- Cloud identification apps provide instant, detailed information about cloud types, helping you confirm your observations and learn more about the clouds you spot.

A. Choosing Cloud Identification Apps

1. CloudSpotter: Developed by the Cloud Appreciation Society, this app helps identify cloud types with a comprehensive catalog of cloud formations and a scoring system to gamify your cloud spotting.

2. GLOBE Observer: This NASA app allows users to contribute their cloud observations to scientific research, providing educational content and tools for identifying clouds.

3. Sky Guide: While primarily an astronomy app, Sky Guide includes features for identifying cloud types and understanding weather patterns.

B. Using Cloud Identification Apps

1. Install and Explore: Download the chosen app and explore its features. Familiarize yourself with the interface and the resources it offers.

2. Identify Clouds: Use the app's camera feature to take a photo of the cloud you're observing. The app will analyze the image and suggest possible cloud types.

3. Learn and Record: Read the detailed descriptions and information provided by the app to enhance your understanding. Record the identified cloud type in your cloud diary or journal.

C. Practical Tips

1. Keep Updated: Regularly update the app to access new features and the latest cloud identification algorithms.

2. Cross-Reference: Use multiple apps to confirm your observations, providing a broader perspective and more accurate identification.

3. Engage in Community: Participate in app-based communities or forums to share your findings and learn from other cloud spotters.

4.2.2 Photography Tips

Purpose:

- Photography allows you to capture and document your cloud observations, creating a visual record that can be analyzed and shared.

A. Camera Settings

1. ISO Settings: Use a low ISO setting (100-200) to reduce noise and ensure clear, sharp images.

2. Aperture: Set a small aperture (f/8 to f/16) for a deep depth of field, ensuring that the entire cloud formation is in focus.

3. Shutter Speed: Adjust the shutter speed based on the movement of the clouds. Use a faster shutter speed (1/250 or faster) to capture moving clouds without blur.

B. Composition Techniques

1. Framing: Include points of interest like trees, buildings, or the horizon to provide scale and context to your cloud photographs.

2. Rule of Thirds: Position the main cloud formation off-center, aligning it with the rule of thirds to create a more dynamic and engaging composition.

3. Foreground Elements: Use elements in the foreground to add depth and perspective to your cloud photos.

C. Lighting and Timing

1. Golden Hour: Photograph clouds during the golden hours (shortly after sunrise or before sunset) to capture dramatic lighting and vibrant colors.

2. Backlighting: Experiment with backlighting to highlight the edges and shapes of clouds, creating a striking contrast between the bright sky and the cloud's silhouette.

3. Weather Conditions: Take advantage of different weather conditions to capture a variety of cloud types and formations.

D. Practical Tips

1. Use a Tripod: For stability, especially in low light or when using a telephoto lens.

2. Review and Edit: Regularly review your photos and edit them to enhance clarity and highlight cloud details.

3. Protect Your Gear: Use weather-sealed bags and lens hoods to protect your equipment from the elements.

4.2.3 Recording Observations

Purpose:

- Recording your observations systematically allows you to track patterns, learn from your experiences, and contribute valuable data to the cloud spotting community.

A. Setting Up Your Cloud Diary

1. Choose a Format: Decide between a physical notebook or a digital journal. Digital journals can include photos and app data, while physical notebooks provide a tactile experience.

2. Sections: Organize your diary into sections for date, time, location, weather conditions, cloud types, and any additional notes.

B. What to Record

1. Basic Information: Record the date, time, and location of your observation.

2. Weather Conditions: Note the temperature, humidity, wind speed, and direction. Use weather apps for precise data.

3. Cloud Types and Descriptions: Identify the cloud types observed and describe their appearance, including color, shape, and movement.

4. Photographs: Attach or link photographs to your entries for visual reference.

C. Consistency and Analysis

1. Regular Entries: Make entries regularly to track changes and patterns over time.

2. Analyze Trends: Periodically review your diary to identify trends and correlations between different cloud types and weather conditions.

3. Share and Compare: Share your findings with cloud spotting communities to compare observations and gain insights from others.

D. Practical Tips

1. Be Detailed: Include as much detail as possible in your entries to create a comprehensive record.

2. Use Apps: Integrate data from cloud identification and weather apps to enhance the accuracy of your recordings.

3. Stay Organized: Keep your diary well-organized for easy reference and analysis.

Practical Application for Advanced Cloud Spotting Techniques

To effectively use these advanced techniques, follow these steps:

1. Preparation:

   - Ensure your apps are installed and up-to-date.

- Gather your camera equipment and review settings.

- Prepare your cloud diary for detailed entries.

Observation:

- Use cloud identification apps to confirm cloud types.

- Take detailed photographs using recommended camera settings and composition techniques.

Documentation:

- Record your observations systematically in your cloud diary.

- Include photos, weather conditions, and detailed descriptions of the clouds.

4. Analysis and Sharing:

- Review your entries regularly to identify patterns and trends.

- Share your observations and photos with cloud spotting communities to gain feedback and insights.

Reflection and Next Steps

Reflect on how using cloud identification apps, photography, and systematic recording can enhance your cloud spotting experience. Consider how these advanced techniques can improve your accuracy and understanding of clouds. Prepare for the next chapter, where we will explore in-depth cloud identification and analysis techniques.

Review Questions

1. How can cloud identification apps enhance your cloud spotting experience?

2. What are some key photography tips for capturing clouds effectively?

3. Why is it important to record your cloud observations systematically?

4. How can you use your recorded observations to improve your cloud spotting skills?

Conclusion

Advanced tools and techniques are essential for mastering cloud spotting. By effectively using cloud identification apps, photography, and systematic recording, you can observe and document clouds in greater detail, enhancing your understanding and appreciation of the sky. Continue to practice and refine your skills, and share your findings with the cloud spotting community. Happy cloud spotting!

4.3 Creating a Cloud Spotting Journal

A cloud spotting journal is a valuable tool for tracking your observations, identifying patterns, and improving your cloud identification skills. This section will guide you through setting up your journal, what to record, and tips for making consistent entries.

4.3.1 Setting Up Your Journal

Purpose:

- A cloud spotting journal helps you systematically document your observations, analyze trends, and enhance your understanding of cloud formations and weather patterns.

A. Choosing the Format

1. Physical Journal: A traditional notebook offers a tactile experience and is easy to carry outdoors. Choose one with durable pages and a weather-resistant cover.

2. Digital Journal: Apps like Evernote, OneNote, or specialized cloud spotting apps allow for multimedia entries, including photos and videos. They offer easy organization and searchability.

3. Hybrid Approach: Combine both methods by using a physical journal for field notes and transferring detailed observations and photos to a digital format.

B. Organizing the Journal

1. Sections: Divide your journal into sections such as date, time, location, weather conditions, cloud types, and notes. This structure helps keep your entries organized and easy to reference.

2. Templates: Create or use pre-made templates to ensure you consistently capture all relevant information. Templates can include fields for weather data, cloud descriptions, and sketches.

C. Essential Supplies

1. Writing Tools: Use pens or pencils that can write clearly on your chosen journal material, including weather-resistant pens for outdoor use.

2. Photography Equipment: Keep a camera or smartphone handy to capture detailed images of cloud formations.

3. Weather Instruments: If possible, use basic weather instruments like a thermometer, hygrometer, and anemometer to record precise weather conditions.

4.3.2 What to Record

Purpose:

- Recording detailed and accurate information in your journal helps you track changes, identify patterns, and make more informed predictions about weather conditions.

A. Basic Information

1. Date and Time: Always start your entry with the date and time to establish a timeline for your observations.

2. Location: Note the specific location where you made your observation. Include latitude and longitude if possible.

B. Weather Conditions

1. Temperature: Record the temperature at the time of your observation. This can be done using a thermometer or checking a weather app.

2. Humidity: Note the humidity level, as it influences cloud formation and characteristics.

3. Wind: Record wind speed and direction. This information helps understand cloud movement and weather patterns.

4. Precipitation: Note any precipitation, such as rain, snow, or hail, and its intensity.

C. Cloud Characteristics

1. Cloud Types: Identify and record the types of clouds observed (e.g., cirrus, cumulus, stratus).

2. Descriptions: Describe the appearance of the clouds, including their shape, color, texture, and any unique features.

3. Altitude: Estimate the altitude of the clouds (high, mid, or low-level) based on their appearance and the weather conditions.

D. Additional Notes

1. Sketches and Photos: Include sketches or photos of the clouds to provide a visual reference for your written descriptions.

2. Weather Changes: Note any significant weather changes that occurred before, during, or after your observation.

3. Personal Observations: Record any thoughts or hypotheses about the weather patterns and cloud formations you observed.

4.3.3 Tips for Consistent Entries

Purpose:

- Consistency in your journal entries helps create a reliable record of your observations, making it easier to identify trends and improve your cloud spotting skills.

A. Set a Routine

1. Regular Observations: Set a regular schedule for your cloud spotting activities. Daily or weekly observations provide a comprehensive view of weather patterns.

2. Specific Times: Choose specific times of the day for your observations, such as morning, midday, and evening. This helps capture how cloud formations change throughout the day.

B. Be Detailed

1. Thorough Descriptions: Take the time to write detailed descriptions of your observations. The more information you include, the more valuable your entries will be for future reference.

2. Use Templates: Use the templates you created to ensure you consistently capture all relevant information.

C. Review and Reflect

1. Regular Review: Periodically review your journal entries to identify trends, patterns, and any gaps in your observations.

2. Reflect on Learnings: Reflect on what you've learned from your entries and consider how you can apply this knowledge to improve your cloud spotting skills.

D. Stay Organized

1. Index and Label: Index your entries and label them clearly to make it easy to find specific observations.

2. Backup Digital Entries: If using a digital journal, regularly back up your entries to avoid losing valuable data.

Practical Application for Creating a Cloud Spotting Journal

To effectively use your cloud spotting journal, follow these steps:

1. Preparation:

   - Choose your journal format (physical, digital, or hybrid) and gather all necessary supplies.

   - Set up your journal with the organized sections and templates.

2. Observation:

- Conduct regular cloud spotting sessions, noting the date, time, location, and weather conditions.

- Identify and describe the clouds you observe, including detailed sketches or photographs.

3. Documentation:

- Record all observations systematically in your journal, ensuring you capture all relevant information.

- Use weather instruments and apps to enhance the accuracy of your entries.

4. Analysis:

- Review your journal entries regularly to identify patterns and trends.

- Reflect on your observations to improve your understanding of cloud formations and weather patterns.

5. Sharing:

- Share your journal entries and insights with cloud spotting communities.

- Engage with other enthusiasts to learn from their experiences and refine your skills.

Reflection and Next Steps

Reflect on the value of keeping a detailed cloud spotting journal. Consider how regular and consistent entries can enhance your understanding of clouds and weather patterns. Prepare for the next chapter, where we will explore in-depth cloud identification and analysis techniques.

Review Questions

1. What are the key elements to include when setting up your cloud spotting journal?

2. Why is it important to record detailed weather conditions and cloud characteristics?

3. How can you ensure consistency in your journal entries?

4. What benefits can you gain from regularly reviewing and reflecting on your journal entries?

Conclusion

Creating and maintaining a cloud spotting journal is an essential practice for serious cloud spotters. By systematically recording your observations, you can track patterns, make informed predictions, and deepen your understanding of cloud formations and weather. Continue to observe, document, and analyze your findings to become a skilled and knowledgeable cloud spotter. Happy cloud spotting!

Chapter 4 Review

4.1 Essential Gear for Cloud Spotting

- Binoculars: Use binoculars to observe cloud details that are not visible to the naked eye. Choose binoculars with 7x to 10x magnification, a wide

- Camera Equipment: Capture and document your cloud observations with a DSLR or mirrorless camera. Use a wide-angle lens for expansive sky scenes and a telephoto lens for specific cloud details. Employ a tripod for stability and set your camera to low ISO, small aperture, and fast shutter speed for sharp images.
- Weather Apps: Use weather apps like AccuWeather, Weather Underground, and CloudSpotter to get real-time data and forecasts. Check forecasts, view satellite imagery, and track radar maps to plan your cloud spotting activities effectively.

4.2 Advanced Techniques for Cloud Spotting

- Cloud Identification Apps: Use apps like CloudSpotter, GLOBE Observer, and Sky Guide to identify cloud types and gain detailed information. These apps analyze cloud photos and provide descriptions to help you confirm your observations.
- Photography Tips: Set your camera to low ISO, small aperture, and fast shutter speed to capture clear cloud images. Use composition techniques like the rule of thirds and include foreground elements for context. Experiment with different lighting conditions, especially during the golden hour.
- Recording Observations: Maintain a cloud spotting journal to systematically document your observations. Record the date, time, location, weather conditions, cloud types, descriptions, and photographs. Use templates to ensure consistency and regularly review your entries to identify trends.

4.3 Creating a Cloud Spotting Journal

- Setting Up Your Journal: Choose between a physical, digital, or hybrid format. Organize your journal with sections for date, time, location, weather conditions, cloud types, and notes. Use templates to ensure comprehensive entries.
- What to Record: Include basic information such as date, time, and location. Record weather conditions like temperature, humidity, wind, and precipitation. Document cloud characteristics including types, descriptions, and altitude. Add additional notes, sketches, and photos.

- Tips for Consistent Entries: Set a regular observation schedule and be thorough in your descriptions. Use templates for consistency and regularly review your entries. Stay organized by indexing and labeling your entries and backing up digital journals.

Practical Application

1. Preparation:

- Gather your binoculars, camera equipment, and weather apps.

- Set up your cloud spotting journal with organized sections and templates.

2. Observation:

- Use binoculars to observe cloud details.

- Capture photographs using recommended settings and techniques.

- Identify clouds using cloud identification apps.

3. Documentation:

- Record observations systematically in your journal.

- Include photos, weather conditions, and detailed cloud descriptions.

4. Analysis:

- Regularly review journal entries to identify patterns and trends.

- Reflect on your observations to improve cloud spotting skills.

5. Sharing:

- Share your findings with cloud spotting communities.

- Engage with other enthusiasts to learn and refine your skills.

Reflection and Next Steps

Reflect on how using essential gear, advanced techniques, and maintaining a cloud spotting journal can enhance your cloud spotting experience. Consider how these practices can improve your accuracy and understanding of clouds. Prepare for the next chapter, where we will explore in-depth cloud identification and analysis techniques.

Review Questions

1. What are the key features to look for when choosing binoculars for cloud spotting?

2. How can camera equipment and photography tips enhance your cloud spotting experience?

3. Why is it important to use cloud identification apps?

4. How can maintaining a cloud spotting journal improve your cloud spotting skills?

Conclusion

Chapter 4 has equipped you with the tools and techniques essential for mastering cloud spotting. By using binoculars, camera equipment, weather apps, cloud identification apps, and maintaining a detailed cloud spotting journal, you can enhance your observation skills and deepen your understanding of cloud formations and weather patterns. Continue to practice, document, and share your findings to become a skilled and knowledgeable cloud spotter. Happy cloud spotting!

Chapter 5: In-Depth Cloud Identification

5.1 Recognizing Shapes, Textures, and Colors

Understanding the shapes, textures, and colors of clouds is fundamental to becoming a proficient cloud spotter. This chapter delves into the specifics of identifying clouds based on these attributes, enhancing your ability to recognize different cloud types and predict weather patterns.

5.1.1 Identifying Features

Clouds can take on a myriad of shapes, each indicative of particular weather conditions. Here, we will explore the most common cloud shapes and their distinguishing features.

A. Cumulus Clouds

- Shape: Puffy, cotton-like appearance with distinct edges.

- Height: Typically form at low altitudes (below 6,500 feet).

- Weather Indication: Generally indicate fair weather but can develop into larger storm clouds.

B. Stratus Clouds

- Shape: Uniform gray layer covering the sky, resembling a blanket.

- Height: Found at low altitudes (up to 6,500 feet).

- Weather Indication: Often associated with overcast skies and light drizzle or mist.

C. Cirrus Clouds

- Shape: Wispy, feather-like streaks high in the sky.

- Height: High altitude clouds (above 20,000 feet).

- Weather Indication: Usually signify fair weather but can precede a change in weather.

D. Nimbus Clouds

- Shape: Thick, dark, and amorphous, often covering the sky.

- Height: Can occur at various altitudes.

- Weather Indication: Associated with continuous, heavy precipitation.

5.1.2 Color Changes

The color of clouds can reveal a lot about their composition and the weather conditions they bring. Understanding these color changes is essential for accurate cloud identification.

A. White Clouds

- Indication: Typical of fair weather. These clouds reflect most of the sunlight, which makes them appear white.

- Common Types: Cumulus and cirrus clouds.

B. Gray Clouds

- Indication: Signify thicker clouds with a higher water content. They can be a precursor to rain or storms.

- Common Types: Stratus and stratocumulus clouds.

C. Dark Gray or Black Clouds

- Indication: Indicate heavy, impending precipitation and storms. The dense water droplets or ice crystals block more sunlight.

- Common Types: Nimbostratus and cumulonimbus clouds.

D. Yellow or Orange Clouds

- Indication: Often observed during sunrise or sunset due to the scattering of sunlight. These colors can also indicate the presence of dust or pollution in the atmosphere.

- Common Types: Can occur with various cloud types, typically during specific times of the day.

E. Red or Pink Clouds

- Indication: Usually seen during sunrise or sunset when the sun is low on the horizon. The long path through the atmosphere scatters the shorter blue wavelengths, leaving the red hues.

- Common Types: Can be seen in cirrus clouds, giving a striking appearance.

5.1.3 Texture Variations

The texture of clouds provides insights into the atmospheric conditions and the processes occurring within the clouds.

A. Smooth and Uniform Texture

- Indication: Often found in stratus clouds, indicating stable air layers and gentle lifting of moist air.

- Example: Stratus clouds creating overcast conditions.

B. Lumpy and Puffy Texture

- Indication: Common in cumulus clouds, indicating pockets of rising warm air and convection processes.

- Example: Fair-weather cumulus clouds on a sunny day.

C. Fibrous and Wispy Texture

- Indication: Seen in cirrus clouds, indicating high-altitude ice crystals and strong upper-level winds.

- Example: Cirrus clouds streaming across the sky in fine strands.

D. Ragged and Frayed Texture

- Indication: Often found in nimbostratus or stratocumulus clouds, suggesting turbulent conditions and possible precipitation.

- Example: Nimbostratus clouds during a rainstorm.

Practical Tips for Cloud Spotting

- A. Observe Regularly: Spend time observing clouds at different times of the day and in various weather conditions to familiarize yourself with their features.

- B. Keep a Journal: Record your observations, noting the cloud types, their shapes, colors, textures, and associated weather conditions.

- C. Use Tools: Utilize cloud charts, mobile apps, and cameras to aid in identifying and documenting different cloud types.

- D. Learn Continuously: Enhance your knowledge by reading meteorology books, attending weather seminars, and participating in cloud-spotting communities.

By mastering the identification of shapes, textures, and colors of clouds, you will gain a deeper understanding of the atmospheric conditions and improve your cloud-spotting skills. This knowledge not only enriches your appreciation of the natural world but also equips you with the ability to anticipate weather changes accurately.

5.2 Seasonal and Regional Variations

Cloud spotting is not just about identifying cloud types but also understanding how they vary with the seasons and regions. This section provides a detailed guide to recognizing these variations and understanding their implications.

5.2.1 Clouds by Season

Different seasons bring distinct weather patterns and cloud formations. Here, we explore the typical clouds associated with each season and their characteristics.

A. Spring

- Common Clouds: Cumulus, Stratocumulus, Cumulonimbus

- Characteristics: Spring is a transition season where warm and cold air masses often clash, leading to the development of cumulus and cumulonimbus clouds. These clouds can bring sudden showers and thunderstorms.

- Weather Indication: Expect frequent changes in weather, with a mix of fair days and stormy conditions.

B. Summer

- Common Clouds: Cumulus, Cumulonimbus

- Characteristics: The heat of summer enhances convection, leading to the formation of large cumulus clouds. On particularly hot days, these can develop into towering cumulonimbus clouds, often resulting in thunderstorms.

- Weather Indication: Summer clouds often signal fair weather, but be alert for afternoon thunderstorms, especially in humid regions.

C. Autumn

- Common Clouds: Stratus, Altostratus, Nimbostratus

- Characteristics: As temperatures drop, stratus and altostratus clouds become more common, bringing overcast skies. Nimbostratus clouds can bring prolonged periods of rain.

- Weather Indication: Expect more consistent overcast conditions with frequent light to moderate rain.

D. Winter

- Common Clouds: Stratus, Nimbostratus, Cirrus

- Characteristics: Cold winter air often leads to the formation of low stratus clouds, creating gray, overcast days. Nimbostratus clouds can bring snow. High cirrus clouds may indicate incoming weather changes.

- Weather Indication: Look for prolonged overcast skies and potential snow, with cirrus clouds hinting at upcoming changes in weather.

5.2.2 Regional Cloud Patterns

Cloud formations can vary significantly from one region to another. Understanding these patterns is crucial for accurate cloud spotting and weather prediction.

A. Coastal Regions

- Common Clouds: Stratus, Fog, Stratocumulus

- Characteristics: Coastal areas often experience stratus clouds and fog due to the interaction of moist ocean air with land. Stratocumulus clouds are also common, particularly in the morning and evening.

- Weather Indication: Expect frequent fog and overcast conditions, with stratocumulus clouds indicating mild weather.

B. Mountainous Regions

- Common Clouds: Cumulus, Orographic Clouds, Lenticular Clouds

- Characteristics: Mountains influence air flow, leading to the formation of orographic clouds and lenticular clouds, which have a lens-like shape. Cumulus clouds are common during the day due to rising warm air.

- Weather Indication: Look for unique cloud formations like lenticular clouds, which can indicate strong winds and turbulence.

C. Plains and Deserts

- Common Clouds: Cumulus, Cumulonimbus, Cirrus

- Characteristics: Wide open spaces allow for significant convection, leading to cumulus and cumulonimbus clouds. Cirrus clouds can often be seen high in the sky.

- Weather Indication: Expect fair weather with cumulus clouds, but be wary of sudden thunderstorms from cumulonimbus clouds.

D. Polar Regions

- Common Clouds: Stratus, Stratocumulus, Nimbostratus

- Characteristics: Cold temperatures lead to low, thick clouds like stratus and stratocumulus. Nimbostratus clouds can bring snow.

- Weather Indication: Anticipate overcast skies and potential snow, with limited sunlight penetration.

5.2.3 Special Weather Events

Certain weather events bring unique cloud formations. Recognizing these special patterns can enhance your cloud spotting expertise.

A. Thunderstorms

- Common Clouds: Cumulonimbus

- Characteristics: Thunderstorms are characterized by towering cumulonimbus clouds, which can reach great heights and produce severe weather.

- Weather Indication: Be prepared for heavy rain, lightning, hail, and potentially tornadoes.

B. Hurricanes

- Common Clouds: Cirrus, Cumulonimbus

- Characteristics: Hurricanes are marked by dense cumulonimbus clouds around the eye and high cirrus clouds spreading outwards.

- Weather Indication: These clouds signal severe weather, with strong winds and heavy rain.

C. Cold Fronts

- Common Clouds: Cumulus, Cumulonimbus, Nimbostratus

- Characteristics: Cold fronts can push warm air upward, leading to cumulus and cumulonimbus clouds. Nimbostratus clouds can bring continuous precipitation.

- Weather Indication: Expect sudden changes in weather, with possible storms and temperature drops.

D. Warm Fronts

- Common Clouds: Stratus, Altostratus, Nimbostratus

- Characteristics: Warm fronts cause gradual lifting of air, leading to layered clouds like stratus and altostratus. Nimbostratus clouds can bring steady rain.

- Weather Indication: Anticipate prolonged periods of precipitation and gradual warming.

Practical Tips for Cloud Spotting

- A. Observe Regularly: Spend time observing clouds at different times of the day and in various weather conditions to familiarize yourself with their features.

- B. Keep a Journal: Record your observations, noting the cloud types, their shapes, colors, textures, and associated weather conditions.

- C. Use Tools: Utilize cloud charts, mobile apps, and cameras to aid in identifying and documenting different cloud types.

- D. Learn Continuously: Enhance your knowledge by reading meteorology books, attending weather seminars, and participating in cloud-spotting communities.

By mastering the identification of seasonal and regional cloud variations, you will gain a deeper understanding of the atmospheric conditions and improve your cloud-spotting skills. This knowledge not only enriches your appreciation of the natural world but also equips you with the ability to anticipate weather changes accurately.

5.3 *Understanding Cloud Dynamics and Formation*

To become an expert cloud spotter, it's crucial to understand the underlying dynamics and processes that lead to cloud formation. This section provides a comprehensive guide to the atmospheric conditions, weather fronts, and the roles of air pressure and temperature in cloud development.

5.3.1 Atmospheric Conditions

Clouds form due to various atmospheric conditions that influence the condensation of water vapor. Understanding these conditions is the first step in recognizing and predicting cloud formations.

A. *Humidity and Moisture Content*

- Role of Humidity: Clouds form when moist air rises and cools, reaching its dew point. High humidity means there's more water vapor available to condense into cloud droplets.

- How to Observe: Use a hygrometer to measure humidity levels. High humidity often correlates with more cloud cover and potential precipitation.

B. Temperature Lapse Rate

- Definition: The rate at which air temperature decreases with altitude. A standard lapse rate is about 6.5°C per kilometer.

- Impact on Clouds: A steep lapse rate (rapid temperature decrease with height) encourages vertical cloud development, such as cumulus and cumulonimbus clouds.

- How to Observe: Pay attention to weather reports and use a thermometer to track temperature changes at different altitudes if possible.

C. Lifting Mechanisms

- Orographic Lift: Occurs when air is forced to rise over mountains. This can lead to cloud formation on the windward side of mountains.

- Convection: Warm air rises due to heating from the ground, leading to cumulus cloud formation.

- Frontal Lift: Happens when air masses with different temperatures collide, causing the warmer air to rise over the cooler air.

- How to Observe: Look for cloud formations near mountains, during warm days and along weather fronts in your region.

5.3.2 Weather Fronts

Weather fronts are boundaries between different air masses and play a significant role in cloud formation and weather patterns. Understanding the types of fronts helps predict the types of clouds and weather that will develop.

A. Cold Fronts

- Characteristics: A cold front occurs when a cold air mass pushes under a warm air mass, forcing the warm air to rise quickly.

- Cloud Formation: Typically results in cumulus and cumulonimbus clouds, leading to thunderstorms and heavy rain.

- How to Observe: Identify a line of rapidly developing, tall clouds, often accompanied by a noticeable drop in temperature.

B. Warm Fronts

- Characteristics: A warm front occurs when a warm air mass slides over a cold air mass, causing gradual lifting of the warm air.

- Cloud Formation: Leads to the formation of stratus, altostratus, and nimbostratus clouds, bringing steady, prolonged precipitation.

- How to Observe: Look for a gradual thickening of clouds from cirrus to stratus, often ahead of the warm front.

C. Stationary Fronts

- Characteristics: Occurs when two air masses are in a standoff, neither strong enough to displace the other.

- Cloud Formation: Can result in prolonged cloudiness and precipitation, with a mix of stratus and nimbostratus clouds.

- How to Observe: Persistent cloudy skies and steady, moderate precipitation are indicators of a stationary front.

D. Occluded Fronts

- Characteristics: Formed when a cold front overtakes a warm front, lifting the warm air off the ground completely.

- Cloud Formation: Can lead to complex cloud structures and varied weather, including cumulus, cumulonimbus, and stratocumulus clouds.

- How to Observe: Mixed cloud types and changing weather conditions, often marked by a significant temperature shift.

5.3.3 Air Pressure and Temperature

Air pressure and temperature are key factors in cloud dynamics. Understanding how these elements interact helps in predicting cloud formation and weather changes.

A. Air Pressure

- High Pressure Systems: Associated with descending air, leading to clear skies and few clouds.

- Low Pressure Systems: Associated with rising air, leading to cloud formation and potential precipitation.

- How to Observe: Use a barometer to measure air pressure. Falling pressure often indicates incoming clouds and weather changes.

B. Temperature

- Role of Temperature: Warmer air can hold more moisture, while cooler air leads to condensation and cloud formation.

- Temperature Inversions: Occur when a layer of warm air traps cooler air below, potentially leading to stratus clouds or fog.

- How to Observe: Track temperature changes using a thermometer and look for signs of inversions, such as persistent low clouds or fog.

C. Dew Point

- Definition: The temperature at which air becomes saturated with moisture and condensation begins.

- Impact on Clouds: When the air temperature drops to the dew point, clouds or fog will form.

- How to Observe: Monitor dew point temperatures using weather instruments or reports to predict cloud formation.

Practical Tips for Cloud Spotting

- A. Observe Regularly: Spend time observing clouds at different times of the day and in various weather conditions to familiarize yourself with their features.

- B. Keep a Journal: Record your observations, noting the cloud types, their shapes, colors, textures, and associated weather conditions.

- C. Use Tools: Utilize cloud charts, mobile apps, and cameras to aid in identifying and documenting different cloud types.

- D. Learn Continuously: Enhance your knowledge by reading meteorology books, attending weather seminars, and participating in cloud-spotting communities.

By mastering the understanding of cloud dynamics and formation, you will gain a deeper insight into the atmospheric conditions that influence weather patterns. This knowledge not only enriches your appreciation of the natural world but also equips you with the ability to anticipate weather changes accurately.

Chapter 5 Review

This chapter covers recognizing shapes, textures, and colors; understanding seasonal and regional variations; and comprehending cloud dynamics and formation.

5.1 Recognizing Shapes, Textures, and Colors

Identifying Features:

- Cumulus: Puffy, cotton-like; indicates fair weather.

- Stratus: Uniform gray layer; associated with overcast skies.

- Cirrus: Wispy, high-altitude streaks; precedes weather changes.

- Nimbus: Thick, dark; signals heavy precipitation.

Color Changes:

- White: Fair weather (Cumulus, Cirrus).

- Gray: Precedes rain (Stratus, Stratocumulus).

- Dark Gray/Black: Heavy rain (Nimbostratus, Cumulonimbus).

- Yellow/Orange: Seen at sunrise/sunset; indicates dust/pollution.

- Red/Pink: Sunrise/sunset hues (Cirrus).

Texture Variations:

- Smooth/Uniform: Stable air layers (Stratus).

- Lumpy/Puffy: Convection (Cumulus).

- Fibrous/Wispy: High-altitude ice crystals (Cirrus).

- Ragged/Frayed: Turbulence, precipitation (Nimbostratus).

5.2 Seasonal and Regional Variations

Clouds by Season:

- Spring: Cumulus, Cumulonimbus (transition season).

- Summer: Cumulus, Cumulonimbus (convection).

- Autumn: Stratus, Altostratus, Nimbostratus (cooling temperatures).

- Winter: Stratus, Nimbostratus, Cirrus (cold air masses).

Regional Cloud Patterns:

- Coastal: Stratus, Fog (moist ocean air).

- Mountainous: Orographic, Lenticular (air flow over mountains).

- Plains/Deserts: Cumulus, Cumulonimbus (convection).

- Polar: Stratus, Nimbostratus (cold temperatures).

Special Weather Events:

- Thunderstorms: Cumulonimbus (severe weather).

- Hurricanes: Cumulonimbus, Cirrus (severe weather).

- Cold Fronts: Cumulus, Nimbostratus (rising warm air).

- Warm Fronts: Stratus, Nimbostratus (gradual lifting).

5.3 Understanding Cloud Dynamics and Formation

Atmospheric Conditions:

- Humidity: High humidity leads to more clouds.

- Temperature Lapse Rate: Steep lapse rate encourages vertical development.

- Lifting Mechanisms: Orographic lift, convection, and frontal lift.

Weather Fronts:

- Cold Fronts: Cumulonimbus, thunderstorms.

- Warm Fronts: Stratus, steady rain.

- Stationary Fronts: Persistent cloudiness.

- Occluded Fronts: Mixed cloud types, complex weather.

Air Pressure and Temperature:

- Air Pressure: High pressure clears skies; low pressure forms clouds.

- Temperature: Warm air holds more moisture; cool air condenses.

- Dew Point: Air temperature drop leads to cloud/fog formation.

Practical Tips for Cloud Spotting

- Observe Regularly: Familiarize with cloud types and conditions.

- Keep a Journal: Record observations.

- Use Tools: Utilize charts, apps, and cameras.

- Learn Continuously: Read meteorology books, attend seminars.

Mastering cloud identification, understanding seasonal and regional variations, and comprehending cloud dynamics will enhance your cloud-spotting skills and weather prediction accuracy.

Part 3: The Science of Clouds

Chapter 6: The Basics of Cloud Formation

Understanding the science behind cloud formation is crucial for effective cloud spotting. This chapter delves into the fundamental processes of the water cycle: evaporation, condensation, and precipitation.

6.1 The Water Cycle: Evaporation, Condensation, Precipitation

The water cycle is a continuous process that involves the movement of water between the Earth's surface and the atmosphere. This cycle is essential for cloud formation and precipitation. Let's explore each stage in detail.

6.1.1 Evaporation

A. Definition and Process

- Definition: Evaporation is the process where water changes from a liquid to a vapor.

- Mechanism: When water on the Earth's surface (oceans, rivers, lakes) absorbs heat from the sun, its molecules gain energy and escape into the air as water vapor.

- Conditions: Higher temperatures and wind increase the rate of evaporation.

B. Observing Evaporation

- Temperature Influence: Warm days with high temperatures accelerate evaporation.

- Wind Impact: Windy conditions enhance evaporation by removing the moist air and replacing it with drier air.

- Surface Area: Larger bodies of water have more surface area for evaporation, making it more significant in those regions.

C. Practical Application

- Cloud Formation: Evaporation is the first step in cloud formation, as it adds water vapor to the atmosphere. Observe water bodies and note the conditions that favor evaporation to predict the likelihood of cloud formation.

6.1.2 Condensation

A. Definition and Process

- Definition: Condensation is the process where water vapor cools and changes back into liquid droplets.

- Mechanism: When warm, moist air rises, it cools. As it cools to its dew point, the water vapor condenses into tiny liquid droplets, forming clouds.

- Conditions: Cooling of air to the dew point is necessary for condensation. This often occurs when air rises and expands in the atmosphere.

B. Observing Condensation

- Dew Point: The dew point is the temperature at which air becomes saturated with moisture and condensation begins. It is crucial for cloud formation.

- Cooling Mechanisms: Air can cool by rising over mountains (orographic lift), meeting a cooler air mass (frontal lift), or expanding as it rises (convective lift).

- Cloud Bases: The base of the cloud forms at the altitude where the rising air reaches its dew point.

C. Practical Application

- Cloud Identification: By understanding where and why condensation occurs, you can identify the type and formation of clouds. Monitor temperature and dew point to predict cloud bases and types.

6.1.3 Precipitation

A. Definition and Process

- Definition: Precipitation is the process where condensed water in clouds falls to the Earth's surface as rain, snow, sleet, or hail.

- Mechanism: Water droplets or ice crystals in clouds combine and grow until they are heavy enough to fall due to gravity. Different temperatures and atmospheric conditions determine the form of precipitation.

- Types: Rain (liquid water), snow (frozen water crystals), sleet (frozen raindrops), hail (layers of ice formed by updrafts in thunderstorms).

B. Observing Precipitation

- Cloud Types: Different cloud types produce different forms of precipitation. For example, nimbostratus clouds bring steady rain, while cumulonimbus clouds can produce heavy rain, hail, or snow.

- Weather Patterns: Study weather patterns and cloud movements to anticipate precipitation events.

- Radar and Satellites: Use weather radar and satellite images to track precipitation in real-time.

C. Practical Application

- Weather Prediction: By understanding precipitation processes and observing cloud types and weather patterns, you can predict impending rain or snow. Note the types of clouds and their development stages to forecast weather conditions accurately.

Practical Tips for Understanding Cloud Formation

- A. Observe Regularly: Spend time observing clouds and weather conditions to understand the processes of evaporation, condensation, and precipitation.

- B. Keep a Journal: Record your observations, noting the conditions that lead to different stages of the water cycle and cloud formation.

- C. Use Tools: Utilize hygrometers, thermometers, and weather charts to measure humidity, temperature, and dew point, which are critical for understanding cloud dynamics.

- D. Learn Continuously: Enhance your knowledge by reading meteorology books, attending weather seminars, and participating in cloud-spotting communities.

By mastering the basics of cloud formation through the water cycle, you will gain a deeper insight into the atmospheric processes that lead to cloud development and precipitation. This knowledge not only enriches your appreciation of the natural world but also equips you with the ability to anticipate weather changes accurately.

6.2 Atmospheric Layers and Their Role in Cloud Formation

Clouds form at different altitudes within the atmosphere. Understanding the characteristics and roles of each atmospheric layer is essential for comprehending how and where various types of clouds develop.

6.2.1 Tropsphere

A. Definition and Characteristics

- Definition: The troposphere is the lowest layer of the Earth's atmosphere, extending from the surface up to about 8-15 kilometers (5-9 miles) depending on latitude and season.

- Characteristics: This layer contains about 75% of the atmosphere's mass and nearly all of its water vapor, which is essential for cloud formation. It is characterized by decreasing temperatures with altitude.

B. Cloud Formation in the Troposphere

- Cloud Types: Most weather phenomena, including cloud formation, occur in the troposphere. Common cloud types include cumulus, stratus, and cumulonimbus.

- Mechanisms: Clouds form due to various processes such as convection (rising warm air), orographic lift (air rising over mountains), and frontal systems (interaction between different air masses).

C. Practical Application

- Weather Prediction: Observing cloud types and behaviors in the troposphere is crucial for short-term weather forecasting. For example, cumulonimbus clouds indicate potential thunderstorms.

- Cloud Spotting: Focus your observations on this layer, as it is where the majority of clouds form. Use binoculars or a camera to capture and study these clouds in detail.

6.2.2 Stratosphere

A. Definition and Characteristics

- Definition: The stratosphere is the atmospheric layer above the troposphere, extending from about 15 to 50 kilometers (9 to 31 miles) above the Earth's surface.

- Characteristics: Unlike the troposphere, temperatures in the stratosphere increase with altitude due to the absorption of ultraviolet radiation by the ozone layer. This layer is more stable and less turbulent.

B. Cloud Formation in the Stratosphere

- Cloud Types: Cloud formation is rare in the stratosphere due to its dry and stable conditions. However, some unique clouds can form, such as nacreous clouds (polar stratospheric clouds).

- Mechanisms: These clouds form at very high altitudes in extremely cold conditions, often seen in polar regions during winter.

C. Practical Application

- Observing Stratospheric Clouds: While rare, spotting nacreous clouds can be a unique experience. Look for these iridescent clouds during the polar winter.

- Impact on Weather: Understanding stratospheric dynamics is essential for comprehending larger-scale atmospheric processes but has less direct impact on everyday weather prediction compared to the troposphere.

6.2.3 Mesosphere

A. Definition and Characteristics

- Definition: The mesosphere is the atmospheric layer above the stratosphere, extending from about 50 to 85 kilometers (31 to 53 miles) above the Earth's surface.

- Characteristics: Temperatures decrease with altitude in the mesosphere, making it the coldest layer of the atmosphere. This layer is less studied due to its inaccessibility.

B. Cloud Formation in the Mesosphere

- Cloud Types: The primary clouds in the mesosphere are noctilucent clouds, also known as polar mesospheric clouds. These are the highest clouds in the Earth's atmosphere, forming at altitudes of around 80 kilometers (50 miles).

- Mechanisms: Noctilucent clouds form when water vapor freezes onto dust particles in the extremely cold temperatures of the mesosphere. They are most visible during twilight in polar regions.

C. Practical Application

- Observing Noctilucent Clouds: These clouds are best viewed during the summer months at high latitudes, just after sunset or before sunrise. They appear as glowing, wispy structures against the darker sky.

- Scientific Importance: While they do not affect daily weather, studying noctilucent clouds can provide insights into atmospheric processes and climate change.

Practical Tips for Cloud Spotting

- A. Understand Layers: Familiarize yourself with the characteristics of each atmospheric layer to better predict where different clouds form.

- B. Use Tools: Utilize telescopes or high-powered binoculars for spotting high-altitude clouds like noctilucent and nacreous clouds.

- C. Record Observations: Keep a detailed log of cloud types, altitudes, and conditions to enhance your understanding of cloud formation across different atmospheric layers.

- D. Continuous Learning: Stay updated with meteorological studies and satellite imagery to track and study cloud formations in various atmospheric layers.

By mastering the roles of the troposphere, stratosphere, and mesosphere in cloud formation, you will deepen your understanding of atmospheric processes. This knowledge not only enriches your cloud spotting experience but also enhances your ability to interpret and predict weather patterns accurately.

6.3 The Impact of Temperature, Pressure, and Humidity

Understanding how temperature, pressure, and humidity interact is essential for cloud spotting. These elements determine cloud formation, types, and weather conditions. This section provides a detailed guide on their roles and how to observe them effectively.

6.3.1 Temperature Variations

A. Definition and Role in Cloud Formation

- Definition: Temperature refers to the degree of heat present in the atmosphere, which affects air density and moisture capacity.

- Role: Warm air holds more moisture, while cold air leads to condensation and cloud formation. Temperature changes drive the vertical movement of air, essential for different cloud types.

B. Observing Temperature Variations

- Daily Fluctuations: Note the temperature changes from day to night. Warm days lead to evaporation, while cooler nights can result in condensation and cloud formation.

- Seasonal Changes: Different seasons bring varying temperatures, influencing cloud types. For example, summer temperatures favor cumulus clouds, while winter promotes stratus clouds.

C. Practical Application

- Use Thermometers: Measure temperature regularly to predict cloud formation. High temperatures during the day often result in cumulus clouds by afternoon.

- Temperature Lapse Rate: Understanding the lapse rate (temperature decrease with altitude) helps in identifying cloud development stages, especially for towering clouds like cumulonimbus.

6.3.2 Air Pressure Differences

A. Definition and Role in Cloud Formation

- Definition: Air pressure is the force exerted by the weight of air in the atmosphere. It decreases with altitude.

- Role: Low pressure areas encourage air to rise, leading to cloud formation, while high pressure areas promote descending air, leading to clear skies.

B. Observing Air Pressure Differences

- Pressure Systems: Identify high and low-pressure systems using weather maps. Low pressure is often associated with cloudy, stormy weather, while high pressure indicates fair weather.

- Barometric Readings: Use a barometer to measure air pressure changes. Falling pressure typically precedes cloud formation and worsening weather.

C. Practical Application

- Weather Prediction: Monitor pressure trends. Rapid drops in pressure can indicate approaching storms and cloud formation.

- Cloud Identification: Recognize the type of clouds associated with different pressure systems, such as cumulonimbus with low pressure and cirrus with high pressure.

6.3.3 Humidity Levels

A. Definition and Role in Cloud Formation

- Definition: Humidity is the amount of water vapor present in the air. Relative humidity is the percentage of moisture the air can hold at a specific temperature.

- Role: High humidity levels increase the likelihood of cloud formation as the air becomes saturated and reaches the dew point, leading to condensation.

B. Observing Humidity Levels

- Hygrometer Use: Measure humidity with a hygrometer. High readings suggest that cloud formation is likely, especially if the temperature drops.

- Dew Point: The temperature at which air becomes saturated and condensation begins. Monitoring dew point helps predict when and where clouds will form.

C. Practical Application

- Daily Observations: Track humidity levels throughout the day. High morning humidity can lead to fog or low stratus clouds, while high afternoon humidity combined with heat can result in cumulus clouds.

- Cloud Development: Use humidity data to understand and predict cloud stages, from formation to precipitation. High humidity near the ground often leads to stratiform clouds, while higher altitude humidity supports cumuliform clouds.

Practical Tips for Cloud Spotting

- A. Regular Monitoring: Observe temperature, pressure, and humidity regularly using reliable instruments. Consistent data helps build an understanding of local weather patterns and cloud behavior.

- B. Keep Detailed Records: Maintain a journal of temperature, pressure, and humidity readings along with observed cloud types and weather conditions. This data is invaluable for recognizing patterns and making accurate predictions.

- C. Combine Observations: Integrate temperature, pressure, and humidity data to form a comprehensive view of atmospheric conditions. Understanding how these elements interact will improve your cloud spotting accuracy.

- D. Use Technology: Leverage weather apps and online resources to access real-time data on temperature, pressure, and humidity. These tools enhance your ability to track and predict cloud formations effectively.

By mastering the impact of temperature, pressure, and humidity on cloud formation, you will significantly enhance your cloud spotting skills. This knowledge enables you to predict weather changes accurately and appreciate the complex dynamics of the atmosphere.

Chapter 6 Review

Chapter 6: The Basics of Cloud Formation provides a comprehensive understanding of the processes and factors essential for cloud formation, focusing on the water cycle, atmospheric layers, and the impact of temperature, pressure, and humidity.

6.1 The Water Cycle: Evaporation, Condensation, Precipitation

The water cycle is a continuous process that involves the movement of water between the Earth's surface and the atmosphere. This section breaks down the three main stages: evaporation, condensation, and precipitation.

- Evaporation: Water from oceans, rivers, and lakes absorbs heat from the sun, causing it to evaporate and add moisture to the atmosphere. Higher temperatures, wind, and larger surface areas of water bodies increase evaporation rates.

- Condensation: As warm, moist air rises and cools to its dew point, it condenses into tiny liquid droplets, forming clouds. This process often occurs when air rises over mountains, meets cooler air masses, or expands as it rises.

- Precipitation: Condensed water in clouds falls to the Earth's surface as rain, snow, sleet, or hail. Different cloud types produce various forms of precipitation, and weather patterns can be studied to anticipate these events.

6.2 Atmospheric Layers and Their Role in Cloud Formation

Clouds form at different altitudes within the atmosphere. This section discusses the characteristics and roles of the troposphere, stratosphere, and mesosphere.

- Troposphere: The lowest layer of the Earth's atmosphere, where most weather phenomena occur. This layer contains most of the atmosphere's mass and water vapor, with temperatures decreasing with altitude. Common cloud types include cumulus, stratus, and cumulonimbus, formed due to convection, orographic lift, and frontal systems.

- Stratosphere: The layer above the troposphere, characterized by increasing temperatures with altitude due to the ozone layer. Cloud formation is rare due to dry and stable conditions, but unique clouds like nacreous clouds can form in extremely cold conditions during winter in polar regions.

- Mesosphere: The layer above the stratosphere, known for its decreasing temperatures with altitude and being the coldest layer. Noctilucent clouds form in this layer at high altitudes, visible during twilight in polar regions.

6.3 The Impact of Temperature, Pressure, and Humidity

Temperature, pressure, and humidity are critical factors in cloud formation. This section explores their interactions and effects.

- Temperature Variations: Temperature affects air density and moisture capacity, driving the vertical movement of air and cloud types. Observing daily and seasonal temperature changes helps predict cloud formation, with warm air holding more moisture and cold air leading to condensation.

- Air Pressure Differences: Air pressure influences rising and descending air currents. Low-pressure areas encourage air to rise, leading to cloud formation, while high-pressure areas promote descending air, leading to clear skies. Monitoring pressure trends with a barometer helps predict storms and cloud formation.

- Humidity Levels: Humidity, the amount of water vapor in the air, increases the likelihood of cloud formation as the air becomes saturated and reaches the dew point. Measuring humidity with a hygrometer and monitoring dew points help predict condensation and cloud formation stages.

Practical Tips for Cloud Spotting

- Regular Monitoring: Observe temperature, pressure, and humidity regularly using reliable instruments to understand local weather patterns and cloud behavior.

- Keep Detailed Records: Maintain a journal of temperature, pressure, and humidity readings along with observed cloud types and weather conditions.

- Combine Observations: Integrate data to form a comprehensive view of atmospheric conditions, improving cloud spotting accuracy.

- Use Technology: Leverage weather apps and online resources for real-time data on temperature, pressure, and humidity.

By mastering the basics of cloud formation through understanding the water cycle, atmospheric layers, and the impact of temperature, pressure, and humidity, you will significantly enhance your cloud spotting skills. This knowledge enables you to predict weather changes accurately and appreciate the complex dynamics of the atmosphere.

Chapter 7: Advanced Meteorology for Cloud Spotters

Cloud spotting is an art that blends scientific knowledge with observational skills. To become a proficient cloud spotter, understanding and using various meteorological instruments is essential. This chapter delves into the primary tools—thermometers, barometers, and anemometers—that are crucial for accurate weather prediction and cloud spotting.

7.1 Meteorological Instruments and Their Uses

Mastering meteorological instruments enhances your ability to predict weather changes and understand cloud formation. This section provides a comprehensive guide on the most essential instruments for cloud spotting.

7.1.1 Thermometers

A. Definition and Role

- Definition: A thermometer is a device used to measure air temperature, which is a crucial factor in cloud formation and weather conditions.

- Role: Temperature influences air density and moisture capacity, which are vital for cloud development. Accurate temperature readings help predict cloud types and weather patterns by indicating the presence of warm or cold fronts.

B. Types of Thermometers

1. Mercury Thermometers

   - Description: Traditional thermometers that use mercury to measure temperature.

   - Pros: Highly accurate.

- Cons: Less common due to the toxicity of mercury.

2. Digital Thermometers

- Description: Electronic devices that provide quick and accurate temperature readings.

- Pros: Easy to use, often include additional features like humidity measurement.

- Cons: Require batteries and can be less durable.

C. Practical Application

1. Placement

- Step 1: Place the thermometer in a shaded, well-ventilated area away from direct sunlight and artificial heat sources to ensure accurate readings.

- Step 2: Mount the thermometer at a height of approximately 1.5 meters (5 feet) above the ground for consistency.

2. Data Collection

- Step 1: Regularly record temperature readings at the same times each day to track daily and seasonal variations.

- Step 2: Compare the recorded data with historical weather patterns to predict cloud formation and type.

3. Understanding Temperature Variations

- Daily Fluctuations: Warm days can lead to increased evaporation and subsequent cloud formation, while cooler nights can cause condensation.

- Seasonal Changes: Different seasons bring varying temperatures that influence cloud types. For example, summer temperatures favor cumulus clouds, while winter promotes stratus clouds.

7.1.2 Barometers

A. Definition and Role

- Definition: A barometer measures air pressure, a key factor that affects weather systems and cloud formation.

- Role: Changes in air pressure can indicate weather changes. Falling pressure suggests worsening weather and rising pressure indicates fair conditions.

B. Types of Barometers

1. Aneroid Barometers

  - Description: Compact and durable, they use a small, flexible metal box called an aneroid cell.

  - Pros: Ideal for field use, no liquid involved.

  - Cons: Can be less accurate than mercury barometers.

2. Mercury Barometers

  - Description: Use mercury to measure air pressure.

  - Pros: Highly accurate.

  - Cons: Fragile, less portable, and involves mercury, which is toxic.

C. Practical Application

1. Monitoring Pressure

- Step 1: Place the barometer indoors at a stable location free from direct sunlight and drafts.

- Step 2: Regularly check and record barometric readings to detect pressure trends.

2. Interpreting Data

- Step 1: Falling pressure indicates rising air, which often leads to cloud formation and potential storms.

- Step 2: Rising pressure indicates descending air, usually leading to clear skies.

3. Integration with Other Data

- Step 1: Combine barometric readings with temperature and humidity data for a comprehensive weather analysis.

- Step 2: Use this integrated data to predict cloud formation and weather changes more accurately.

7.1.3 Anemometers

A. Definition and Role

- Definition: An anemometer measures wind speed and direction, critical for understanding weather dynamics and cloud movement.

- Role: Wind patterns influence cloud formation and dispersion, helping to predict weather changes and cloud behavior.

B. Types of Anemometers

1. Cup Anemometers

- Description: Use rotating cups to measure wind speed.

- Pros: Simple and reliable.

- Cons: Measures only wind speed, not direction.

2. Vane Anemometers

- Description: Measure both wind speed and direction using a vane that aligns with the wind.

- Pros: Provide a complete picture of wind patterns.

- Cons: More complex than cup anemometers.

C. Practical Application

1. Wind Speed Recording

- Step 1: Install the anemometer in an open area, away from obstructions that could affect wind flow.

- Step 2: Regularly measure and record wind speed to track changes.

2. Wind Direction Analysis

- Step 1: Use a vane anemometer or add a wind vane to a cup anemometer to measure wind direction.

- Step 2: Record wind direction data to understand the movement of weather systems and associated clouds.

3. Combining Wind Data with Other Observations

   - Step 1: Use wind speed and direction data in conjunction with temperature, pressure, and humidity readings to predict cloud movement and weather changes.

   - Step 2: Monitor changes in wind patterns, as sudden shifts can indicate approaching weather fronts or storm systems.

Practical Tips for Using Meteorological Instruments

- A. Regular Monitoring: Consistently use thermometers, barometers, and anemometers to gather accurate data on temperature, pressure, and wind.

- B. Keep Detailed Records: Maintain a weather journal with regular entries on readings from your instruments and corresponding weather conditions.

- C. Integrate Data: Combine readings from different instruments to form a comprehensive understanding of atmospheric conditions.

- D. Use Technology: Utilize weather apps and online tools to complement your data with real-time weather information.

By mastering the use of meteorological instruments, you will significantly enhance your ability to predict weather changes and accurately spot and understand clouds. This knowledge is essential for becoming an expert cloud spotter, enabling you to appreciate the complex dynamics of the atmosphere and make informed weather predictions.

7.2 Satellite Imagery and Doppler Radar

Mastering the use of satellite imagery and Doppler radar is crucial for advanced cloud spotting. This section provides a comprehensive guide on understanding satellite images, using Doppler radar, and interpreting data for accurate weather forecasting.

7.2.1 Understanding Satellite Images

A. Definition and Role

- Definition: Satellite images are photographs of Earth taken by satellites in space, providing a comprehensive view of cloud cover, weather systems, and atmospheric conditions.

- Role: These images help track large-scale weather patterns, monitor cloud movements, and predict weather changes.

B. Types of Satellite Images

1. Visible Light Images

   - Description: These images resemble regular photographs, showing clouds, land, and sea in natural colors.

   - Usage: Best used during daylight hours to observe cloud cover and weather systems.

2. Infrared Images

   - Description: These images capture temperature differences, showing clouds and weather systems based on heat emitted.

   - Usage: Useful for monitoring weather patterns both day and night, as they can indicate cloud height and temperature.

3. Water Vapor Images

   - Description: These images show the distribution of water vapor in the atmosphere.

- Usage: Essential for tracking moisture levels and predicting cloud formation and precipitation.

C. Practical Application

1. Analyzing Cloud Cover

- Step 1: Access satellite imagery from reliable sources such as weather websites or apps.

- Step 2: Observe the distribution and movement of clouds over your area.

- Step 3: Note the cloud types and their formation patterns to predict weather changes.

2. Monitoring Weather Systems

- Step 1: Identify large-scale weather systems such as fronts, storms, and cyclones.

- Step 2: Track the movement and development of these systems to anticipate their impact on local weather.

3. Using Time-Lapse Imagery

- Step 1: Review time-lapse satellite images to observe the progression of cloud cover and weather systems.

- Step 2: Use this information to understand the dynamics of cloud formation and movement over time.

7.2.2 Using Doppler Radar

A. Definition and Role

- Definition: Doppler radar is a specialized radar that uses the Doppler effect to measure the velocity of precipitation particles, providing detailed information on weather conditions.

- Role: Doppler radar helps detect severe weather phenomena like thunderstorms, tornadoes, and heavy precipitation by analyzing precipitation patterns and movements.

B. Key Features of Doppler Radar

1. Reflectivity

- Description: Measures the intensity of precipitation, with higher values indicating heavier rainfall.

- Usage: Helps identify areas of heavy precipitation and potential flooding.

2. Velocity

- Description: Measures the speed and direction of wind within weather systems.

- Usage: Essential for detecting rotation in thunderstorms, indicating possible tornado formation.

3. Dual-Polarization

- Description: Provides additional information on precipitation type by sending out pulses in both horizontal and vertical orientations.

- Usage: Distinguishes between rain, snow, sleet, and hail, improving precipitation forecasts.

C. Practical Application

1. Severe Weather Detection

- Step 1: Access Doppler radar data from reliable sources such as weather websites or apps.

- Step 2: Monitor reflectivity to identify areas of heavy precipitation and potential severe weather.

- Step 3: Check velocity data to detect wind patterns and potential tornado formation.

2. Precipitation Monitoring

- Step 1: Use dual-polarization data to determine the type of precipitation.

- Step 2: Track precipitation intensity and movement to predict rainfall, snowfall, or hail events.

3. Storm Tracking

- Step 1: Observe the development and movement of storms using Doppler radar images.

- Step 2: Monitor changes in reflectivity and velocity to anticipate storm behavior and potential impacts.

7.2.3 Interpreting Data

A. Combining Satellite and Radar Data

- Integrated Analysis: Use both satellite and radar data to get a comprehensive understanding of weather systems and their impact on cloud formation.

B. Practical Applications

1. Weather Prediction

- Step 1: Combine satellite imagery with Doppler radar data to monitor weather systems.

- Step 2: Use integrated data to predict weather changes accurately, including precipitation and storm development.

2. Cloud Identification

- Step 1: Use satellite images to identify cloud types and their formation patterns.

- Step 2: Use Doppler radar data to understand the dynamics within clouds, such as precipitation and wind patterns.

3. Severe Weather Preparedness

- Step 1: Regularly monitor satellite and radar data to stay informed about approaching severe weather.

- Step 2: Use this information to prepare for weather events, ensuring safety and readiness.

Practical Tips for Using Satellite Imagery and Doppler Radar

- A. Regular Monitoring: Consistently use satellite imagery and Doppler radar to gather accurate data on weather systems.

- B. Keep Detailed Records: Maintain a weather journal with entries on satellite and radar observations along with corresponding weather conditions.

- C. Integrate Data: Combine satellite and radar data to form a comprehensive understanding of atmospheric conditions.

- D. Use Technology: Leverage weather apps and online tools to complement your data with real-time weather information.

By mastering the use of satellite imagery and Doppler radar, you will significantly enhance your ability to predict weather changes and accurately spot and understand clouds. This knowledge is essential for becoming an expert cloud spotter, enabling you to appreciate the complex dynamics of the atmosphere and make informed weather predictions.

7.3 Predicting Weather Through Clouds

Predicting weather through clouds involves recognizing specific cloud patterns, understanding their implications, and applying this knowledge to forecast weather conditions. This section provides a detailed, step-by-step guide to using clouds for weather prediction.

7.3.1 Cloud Patterns

Understanding and identifying cloud patterns is essential for predicting weather changes. Different clouds indicate specific weather conditions. Here's how to recognize and interpret these patterns:

A. Definition and Importance of Cloud Patterns

- Definition: Cloud patterns refer to the arrangement, appearance, and behavior of clouds in the sky, which can provide clues about upcoming weather.

- Importance: Recognizing cloud patterns helps in predicting weather changes, from fair conditions to severe storms.

B. Types of Cloud Patterns and Their Implications

1. Cumulus Clouds

- Description: Fluffy, white clouds with a flat base.

- Weather Indication: Typically indicate fair weather, but can develop into cumulonimbus clouds, signaling thunderstorms.

2. Stratus Clouds

- Description: Uniform, gray clouds covering the sky like a blanket.

- Weather Indication: Often associated with overcast conditions and light precipitation such as drizzle.

3. Cirrus Clouds

- Description: High-altitude, wispy clouds made of ice crystals.

- Weather Indication: Indicate a change in weather, often preceding a warm front and possible precipitation within 24 to 48 hours.

4. Cumulonimbus Clouds

- Description: Towering clouds with a flat, anvil-shaped top.

- Weather Indication: Signal severe weather conditions, including thunderstorms, heavy rain, hail, and possible tornadoes.

5. Altostratus Clouds

- Description: Gray or blue-gray clouds that cover the sky, usually at mid-altitude.

- Weather Indication: Often precede storms with continuous rain or snow.

C. Practical Steps to Recognize Cloud Patterns

1. Observation

- Step 1: Regularly observe the sky and note the cloud types present.

- Step 2: Look for changes in cloud patterns throughout the day.

2. Documentation

- Step 1: Keep a cloud journal, recording the types of clouds, their appearance, and associated weather conditions.

- Step 2: Take photographs to compare with future observations and improve pattern recognition skills.

3. Analysis

- Step 1: Analyze the recorded data to identify trends and patterns.

- Step 2: Use this information to predict weather changes based on observed cloud patterns.

7.3.2 Weather Forecasting

Using cloud observations for weather forecasting involves understanding the relationships between cloud types and weather systems. Here's how to use this knowledge for accurate predictions:

A. Role of Clouds in Weather Forecasting

- Indicators: Different clouds indicate specific weather conditions, such as fair weather, precipitation, or storms.

- Forecasting Techniques: By observing cloud types and their development, you can anticipate changes in weather.

B. Practical Steps for Weather Forecasting Through Clouds

1. Short-Term Forecasting

   - Step 1: Observe the sky for immediate cloud patterns and changes.

   - Step 2: Identify the cloud types and their current development stage.

   - Step 3: Use this information to predict short-term weather changes, such as rain or thunderstorms within the next few hours.

2. Long-Term Forecasting

   - Step 1: Monitor cloud patterns over several days.

   - Step 2: Identify trends, such as the arrival of cirrus clouds indicating an approaching warm front.

   - Step 3: Use this data to predict longer-term weather changes, such as rain or snow within the next 24 to 48 hours.

3. Seasonal Forecasting

   - Step 1: Recognize seasonal cloud patterns that recur annually.

   - Step 2: Use historical data and current observations to anticipate seasonal weather changes, such as monsoon rains or winter snowstorms.

7.3.3 Practical Applications

Applying cloud pattern recognition and weather forecasting techniques can improve your daily weather predictions and preparedness. Here's how to put this knowledge into practice:

A. Daily Weather Prediction

1. Morning Observations

   - Step 1: Observe the sky first thing in the morning.

   - Step 2: Identify any significant cloud patterns and their implications for the day's weather.

   - Step 3: Plan your day based on the predicted weather, such as carrying an umbrella if rain is likely.

2. Ongoing Monitoring

   - Step 1: Continue to observe the sky throughout the day.

   - Step 2: Note any changes in cloud patterns and adjust your predictions accordingly.

   - Step 3: Use real-time data from weather apps to complement your observations.

B. Preparing for Severe Weather

1. Early Detection

   - Step 1: Look for early signs of severe weather, such as the development of cumulonimbus clouds.

- Step 2: Monitor weather forecasts and warnings from reliable sources.

   - Step 3: Take proactive measures, such as securing outdoor items and preparing an emergency kit.

2. Real-Time Response

   - Step 1: Stay updated with real-time weather information through radar and satellite images.

   - Step 2: Respond quickly to changes, such as seeking shelter if a storm is imminent.

   - Step 3: Communicate with others to ensure everyone is aware and prepared for severe weather conditions.

C. Enhancing Cloud Spotting Skills

1. Continuous Learning

   - Step 1: Participate in cloud spotting communities and forums to learn from other enthusiasts.

   - Step 2: Attend meteorology workshops and seminars to deepen your understanding.

   - Step 3: Stay updated with the latest research and advancements in meteorology.

2. Practical Exercises

   - Step 1: Conduct regular cloud spotting exercises, documenting and analyzing your observations.

   - Step 2: Compare your predictions with actual weather outcomes to improve accuracy.

- Step 3: Experiment with different observation techniques and tools to find what works best for you.

By mastering the art of predicting weather through clouds, you will significantly enhance your cloud spotting skills and ability to forecast weather changes. This knowledge is essential for becoming an expert cloud spotter, enabling you to appreciate the complex dynamics of the atmosphere and make informed weather predictions.

Chapter 7 Review

Chapter 7: Advanced Meteorology for Cloud Spotters focuses on the tools and techniques essential for accurate cloud spotting and weather prediction. This chapter covers the use of meteorological instruments, satellite imagery, Doppler radar, and predicting weather through clouds.

7.1 Meteorological Instruments and Their Uses

Thermometers: Measure air temperature, influencing cloud formation. Types include mercury and digital thermometers. Place in shaded, well-ventilated areas, and record readings regularly to track variations.

Barometers: Measure air pressure, affecting weather systems. Types include aneroid and mercury barometers. Monitor pressure trends to predict weather changes, combining readings with temperature and humidity data.

Anemometers: Measure wind speed and direction, crucial for understanding weather dynamics. Types include cup and vane anemometers. Record wind speed and direction to predict cloud movement and weather changes.

7.2 Satellite Imagery and Doppler Radar

Understanding Satellite Images: Provide a comprehensive view of cloud cover and weather systems. Types include visible light, infrared, and water vapor images. Analyze these images to monitor cloud patterns and weather systems.

Using Doppler Radar: Measures precipitation particles' velocity, providing detailed weather information. Key features include reflectivity, velocity, and dual-polarization. Use radar data to detect severe weather, monitor precipitation, and track storms.

Interpreting Data: Combine satellite and radar data for a comprehensive understanding of weather systems. Use integrated data to predict weather changes, identify cloud types, and prepare for severe weather.

7.3 Predicting Weather Through Clouds

Cloud Patterns: Recognize specific cloud patterns to predict weather changes. Common patterns include cumulus, stratus, cirrus, cumulonimbus, and altostratus clouds. Observe and document these patterns to anticipate weather conditions.

Weather Forecasting: Use cloud observations for short-term, long-term, and seasonal weather forecasting. Monitor cloud development and trends to predict weather changes accurately.

Practical Applications: Apply cloud pattern recognition to daily weather predictions and severe weather preparedness. Enhance your cloud spotting skills through continuous learning, practical exercises, and participation in cloud spotting communities.

By mastering the use of meteorological instruments, satellite imagery, Doppler radar, and cloud pattern recognition, you will significantly enhance your cloud spotting skills and ability to predict weather changes. This knowledge is essential for becoming an expert cloud spotter, enabling you to appreciate the complex dynamics of the atmosphere and make informed weather predictions.

Part 4: Cloud Spotting as a Lifestyle

Chapter 8: Starting Your Cloud Spotting Journey

Embarking on your cloud spotting journey is an exciting and rewarding endeavor. To make the most of your experience, it's important to understand the best times and places for cloud spotting. This chapter will guide you through identifying optimal weather conditions, choosing ideal locations, and considering seasonal factors for effective cloud spotting.

8.1 Best Times and Places for Cloud Spotting

Finding the right time and place is crucial for successful cloud spotting. This section outlines the key factors to consider.

8.1.1 Optimal Weather Conditions

A. Clear Visibility

- Step 1: Choose days with clear visibility. Avoid cloud spotting during heavy fog, dense clouds, or severe weather conditions that can obstruct your view.

- Step 2: Early mornings and late afternoons are often the best times for cloud spotting due to the softer light, which can highlight cloud details more effectively.

B. Light Winds

- Step 1: Opt for days with light to moderate winds. Strong winds can disrupt cloud formations and make observation challenging.

- Step 2: Monitor local weather reports for wind conditions before heading out.

C. Calm Weather

- Step 1: Avoid extreme weather conditions like thunderstorms, heavy rain, and snow, as they can be dangerous and obscure cloud visibility.

- Step 2: Plan your cloud spotting activities for calm, stable weather days to ensure a safe and enjoyable experience.

8.1.2 Ideal Locations

A. Open Areas

- Step 1: Select open areas with an unobstructed view of the sky, such as parks, fields, or hilltops. These locations provide the best vantage points for observing a wide expanse of the sky.

- Step 2: Avoid urban areas with tall buildings or dense forests that can block your view and limit your cloud spotting experience.

B. High Altitudes

- Step 1: Seek out higher elevations like mountains or hills. These locations offer clearer views of cloud formations and bring you closer to certain cloud types.

- Step 2: Be prepared for varying weather conditions and dress appropriately for higher altitudes.

C. Water Bodies

- Step 1: Consider locations near lakes, rivers, or oceans. Water bodies can enhance cloud formations due to the availability of moisture and often provide stunning reflections.

- Step 2: Ensure you have a safe and accessible spot near the water for cloud spotting.

8.1.3 Seasonal Considerations

A. Spring and Summer

- Step 1: Look for cumulus and cumulonimbus clouds, which are more common during warmer months. These clouds can develop into thunderstorms, providing dynamic cloud spotting opportunities.

- Step 2: Monitor the sky during the afternoon when these clouds are most likely to form and evolve.

B. Fall

- Step 1: Observe the transition in cloud types as the weather cools, including stratus and altostratus clouds. These clouds often indicate changing weather patterns.

- Step 2: Early mornings in fall may offer mist and fog, providing unique opportunities to see lower-level clouds.

C. Winter

- Step 1: Focus on cirrus and stratocumulus clouds, which are common during the colder months. The crisp winter air often leads to clearer skies and more distinct cloud formations.

- Step 2: Be prepared for rapid weather changes and dress warmly to stay comfortable during your cloud spotting activities.

By considering these factors, you can optimize your cloud spotting experience and enjoy the beauty and variety of clouds throughout the year. Whether you're a solo spotter or enjoy sharing the experience with others, knowing the best times and places will enhance your appreciation and understanding of the sky's ever-changing canvas.

8.2 Cloud Spotting Alone vs. in Groups

Cloud spotting can be a solitary activity or a social one, each with its unique benefits and considerations. This section explores the advantages of both solo and group cloud spotting and provides guidance on organizing cloud spotting events.

8.2.1 Benefits of Solo Cloud Spotting

A. Personal Reflection

- Step 1: Use solo cloud spotting as a time for personal reflection and relaxation. The solitude allows you to immerse yourself fully in the experience.

- Step 2: Bring a journal to note your thoughts and observations, fostering a deeper connection with nature and your surroundings.

B. Flexible Schedule

- Step 1: Spot clouds whenever you have free time without the need to coordinate with others. This flexibility makes it easier to take advantage of spontaneous opportunities.

- Step 2: Choose locations and times that suit your preferences, whether it's an early morning hike or a quiet afternoon in the park.

C. Enhanced Focus

- Step 1: Concentrate on observing and learning about clouds at your own pace. Without distractions, you can delve deeply into the details of cloud formations and weather patterns.

- Step 2: Take detailed notes and photographs without interruptions, helping to improve your skills and knowledge over time.

8.22 Advantages of Group Cloud Spotting

A. Social Interaction

- Step 1: Enjoy the camaraderie of sharing your cloud spotting experiences with friends, family, or fellow enthusiasts. Social interactions can make the experience more enjoyable and memorable.

- Step 2: Use group outings to bond over a shared interest, fostering friendships and connections with others who appreciate the beauty of clouds.

B. Diverse Perspectives

- Step 1: Benefit from different viewpoints and interpretations of cloud formations. Each person might notice something you missed, providing a more comprehensive understanding of the sky.

- Step 2: Engage in discussions that deepen your knowledge of meteorology and cloud types, learning from each other's insights and experiences.

C. Organized Activities

- Step 1: Participate in planned activities like cloud identification challenges, photography contests, or educational talks. Structured activities can enhance the learning experience and make it more interactive.

- Step 2: Take advantage of group resources, such as shared equipment, reference materials, and pooled knowledge, to enhance your cloud spotting experience.

8.2.3 Organizing Cloud Spotting Events

A. Planning the Event

- Step 1: Choose a date and time with optimal weather conditions for cloud spotting. Check the weather forecast in advance to ensure good visibility and safe conditions.

- Step 2: Select a location that offers good visibility and comfort for participants. Consider accessibility, parking, and amenities like restrooms and picnic areas.

B. Communicating with Participants

- Step 1: Use social media, community boards, or cloud spotting forums to invite participants. Create an event page with all the necessary details.

- Step 2: Provide information about the meeting point, schedule, and any necessary equipment or preparations. Encourage participants to bring items like binoculars, cameras, notebooks, and weather-appropriate clothing.

C. Conducting the Event

- Step 1: Start with a brief introduction about the day's objectives and expected cloud formations. Provide a basic overview of the types of clouds participants might see and their significance.

- Step 2: Encourage participants to share their observations and knowledge during the event. Foster an inclusive environment where everyone feels comfortable contributing.

- Step 3: Conclude with a discussion or Q&A session to consolidate learning and experiences. Encourage participants to share their favorite sightings and what they learned from the experience.

By understanding the benefits of both solo and group cloud spotting and knowing how to organize successful events, you can enrich your cloud spotting experience. Whether you prefer the solitude of solo spotting or the camaraderie of group outings, these strategies will help you make the most of your cloud spotting adventures.

8.3 Staying Safe While Cloud Spotting

Safety should always be a priority when engaging in cloud spotting. This section provides comprehensive guidelines on how to stay safe while enjoying your cloud spotting activities, including weather safety tips, environmental awareness, and personal safety precautions.

8.3.1 Weather Safety Tips

A. Monitor Weather Forecasts

- Step 1: Check the weather forecast before heading out to ensure safe conditions for cloud spotting. Reliable sources include local weather stations, weather apps, and online meteorological services.

- Step 2: Be aware of any severe weather alerts or warnings. Avoid cloud spotting during thunderstorms, high winds, or other hazardous weather conditions.

B. Avoid Dangerous Conditions

- Step 1: Refrain from cloud spotting during extreme weather conditions such as thunderstorms, heavy rain, or snowstorms, as these can be dangerous and obscure cloud visibility.

- Step 2: If you encounter severe weather while cloud spotting, seek shelter immediately. Safety should always come first.

C. Dress Appropriately

- Step 1: Wear suitable clothing for the weather conditions. In colder months, dress in layers to stay warm and dry. In warmer weather, wear lightweight, breathable fabrics.

- Step 2: Use sun protection, such as hats, sunglasses, and sunscreen, during sunny days to prevent sunburn and heat-related issues.

8.3.2 Environmental Awareness

A. Respect Nature

- Step 1: Practice Leave No Trace principles. Carry out all trash and avoid disturbing wildlife. Leave natural environments as you found them.

- Step 2: Stick to designated paths and trails to minimize environmental impact. Avoid trampling on vegetation and disturbing natural habitats.

B. Stay Hydrated

- Step 1: Bring plenty of water, especially on hot days, to stay hydrated during your cloud spotting activities.

- Step 2: Take regular breaks to rest and hydrate, particularly if you are hiking or in an area with limited shade.

C. Be Aware of Your Surroundings

- Step 1: Be mindful of your footing, especially in rugged terrain or near water bodies. Avoid areas with steep drops or unstable ground.

- Step 2: Keep an eye on changing weather conditions and be prepared to leave if necessary. Sudden changes in weather can occur, and it's important to stay alert.

8.3.3 Personal Safety Precautions

A. Inform Someone of Your Plans

- Step 1: Let a friend or family member know your cloud spotting plans, including your location and expected return time. This ensures someone is aware of your whereabouts in case of an emergency.

- Step 2: Carry a fully charged phone for emergencies. Ensure you have reception in the area you plan to visit or know where the nearest areas with reception are located.

B. Carry Safety Gear

- Step 1: Pack a basic first aid kit, including bandages, antiseptic wipes, and any personal medications you might need. Be prepared for minor injuries or emergencies.

- Step 2: Bring additional safety items such as a flashlight, multi-tool, whistle, or personal alarm. These can be crucial in case of an unexpected situation.

C. Stay Aware of Local Wildlife

- Step 1: Know what wildlife is common in the area and how to stay safe around them. Avoid approaching or feeding wild animals.

- Step 2: Be cautious of insects, snakes, and other potentially harmful wildlife. Wear appropriate clothing and use insect repellent if necessary.

D. Stay Alert and Mindful

- Step 1: Pay attention to your surroundings and avoid distractions, especially if you are in a remote or unfamiliar area.

- Step 2: Trust your instincts. If something feels off or unsafe, leave the area and find a safer location for cloud spotting.

By following these comprehensive safety guidelines, you can ensure a safe and enjoyable cloud spotting experience. Whether you are cloud spotting alone or with a group, these tips will help you stay prepared and protected while you explore and appreciate the beauty of the sky.

Chapter 8 Review

Chapter 8: Starting Your Cloud Spotting Journey offers essential guidelines and practical advice for embarking on your cloud spotting adventure. This chapter covers the best times and places for cloud spotting, compares the benefits of solo and group cloud spotting, and emphasizes the importance of staying safe while engaging in this activity.

8.1 Best Times and Places for Cloud Spotting

Finding the right time and place is crucial for a successful cloud spotting experience.

- *Optimal Weather Conditions:*

   - Clear Visibility: Choose days with good visibility, avoiding heavy fog or severe weather.

   - Light Winds: Prefer days with light to moderate winds to prevent cloud disruption.

   - Calm Weather: Avoid extreme weather conditions for safety and better observation.

- *Ideal Locations:*

   - Open Areas: Select parks, fields, or hilltops with unobstructed views of the sky.

- High Altitudes: Consider mountains or hills for clearer views and closer proximity to certain cloud types.

- Water Bodies: Lakes, rivers, or oceans can enhance cloud formations and provide beautiful reflections.

- Seasonal Considerations:

- Spring and Summer: Look for cumulus and cumulonimbus clouds, especially in the afternoon.

- Fall: Observe transitions to stratus and altostratus clouds, with unique morning fog opportunities.

- Winter: Focus on cirrus and stratocumulus clouds, and be prepared for rapid weather changes.

8.2 Cloud Spotting Alone vs. in Groups

Deciding whether to cloud spot alone or in a group can affect your experience.

- Benefits of Solo Cloud Spotting:

- Personal Reflection: Use the time for solitude and connection with nature.

- Flexible Schedule: Spot clouds whenever convenient without needing to coordinate with others.

- Enhanced Focus: Concentrate on your observations and learn at your own pace.

- Advantages of Group Cloud Spotting:

- Social Interaction: Share experiences and enjoy the camaraderie of fellow enthusiasts.

- Diverse Perspectives: Gain insights from different viewpoints and deepen your knowledge through discussions.

- Organized Activities: Participate in structured events like cloud identification challenges and educational talks.

- Organizing Cloud Spotting Events:

- Planning: Choose optimal weather and accessible locations for the event.

- Communicating: Use social media and forums to invite participants and provide necessary details.

- Conducting: Start with an introduction, encourage sharing of observations, and conclude with a discussion or Q&A session.

8.3 Staying Safe While Cloud Spotting

Safety is paramount when engaging in cloud spotting activities.

- Weather Safety Tips:

- Monitor Forecasts: Check weather reports to ensure safe conditions.

- Avoid Dangerous Conditions: Refrain from spotting during severe weather and seek shelter if needed.

- Dress Appropriately: Wear suitable clothing for the weather, including sun protection on sunny days.

- Environmental Awareness:

- Respect Nature: Practice Leave No Trace principles and avoid disturbing wildlife.

- Stay Hydrated: Bring water and take regular breaks, especially in hot weather.

- Be Aware: Be mindful of your footing and changing weather conditions.

- *Personal Safety Precautions:*

   - Inform Someone: Let a friend or family member know your plans and expected return time.

   - Carry Safety Gear: Pack a first aid kit, flashlight, multi-tool, and other essential items.

   - Stay Aware of Wildlife: Know the local wildlife and take precautions to stay safe.

By understanding the best times and places for cloud spotting, weighing the benefits of solo versus group activities, and prioritizing safety, you can start your cloud spotting journey with confidence and enjoyment. Whether you prefer the tranquility of solo spotting or the shared experience of group outings, these guidelines will help you make the most of your cloud spotting adventures.

Chapter 9: Capturing the Beauty of Clouds

Capturing the beauty of clouds is a rewarding aspect of cloud spotting, allowing you to document and share the stunning formations you observe. This chapter provides comprehensive, step-by-step guidance on photographing clouds, covering essential camera settings, composition techniques, and editing tips.

9.1 Photography Tips and Tricks

9.1.1 Camera Settings

Mastering your camera settings is crucial for capturing high-quality cloud photos. Here's a systematic approach to getting the best results.

A. Choosing the Right Camera

- Step 1: Use a DSLR or mirrorless camera for the best results. These cameras offer superior control over settings and higher image quality.

- Step 2: Modern smartphones with advanced cameras can also produce excellent cloud photos, especially if they offer manual control settings.

B. Optimal Settings

- Step 1: ISO: Set your ISO to 100-200 to minimize noise in your images. This is especially important for maintaining the clarity of cloud details.

- Step 2: Aperture: Use a small aperture (f/8 to f/16) to achieve a deep depth of field, ensuring all parts of the cloud are in focus.

- Step 3: Shutter Speed: Adjust the shutter speed based on lighting conditions. Faster speeds (1/250s or higher) are ideal for capturing moving clouds without blur.

- Step 4: White Balance: Set the white balance to daylight to accurately capture the colors of the sky and clouds. Adjust manually if needed to suit specific lighting conditions.

C. Using Filters

- Step 1: Use a polarizing filter to reduce glare and enhance the contrast and color saturation of clouds.

- Step 2: Consider a graduated neutral density (ND) filter to balance the exposure between the bright sky and the darker landscape.

9.1.2 Composition Techniques

Composition is key to creating visually appealing and dynamic cloud photographs. Follow these techniques to enhance your images.

A. Rule of Thirds

- Step 1: Divide your frame into nine equal sections using two horizontal and two vertical lines (the rule of thirds grid).

- Step 2: Place key elements, such as striking cloud formations or the horizon, along these lines or at their intersections to create a balanced and engaging composition.

B. Leading Lines

- Step 1: Use natural lines, such as roads, rivers, or the horizon, to lead the viewer's eye towards the clouds.

- Step 2: Ensure these lines guide the viewer through the photo, creating a sense of depth and perspective.

C. Foreground Interest

- Step 1: Include elements in the foreground, such as trees, buildings, or mountains, to add context and scale to your cloud photos.

- Step 2: Ensure the foreground elements complement the clouds and do not overpower the main subject.

D. Framing

- Step 1: Use natural frames, like tree branches or archways, to highlight and draw attention to the clouds.

- Step 2: Ensure the frame does not obstruct the view but enhances the focus on the clouds.

9.1.3 Editing Tips

Editing your photos can significantly enhance their quality and impact. Follow these tips for effective photo editing.

A. Basic Adjustments

- Step 1: Use editing software like Adobe Lightroom or Photoshop to adjust exposure, contrast, and saturation.

- Step 2: Enhance details by adjusting shadows and highlights to bring out the textures and nuances of the clouds.

B. Advanced Techniques

- Step 1: Use the clarity and dehaze sliders to increase the contrast and definition of the clouds, making them stand out.

- Step 2: Apply selective editing to emphasize certain parts of the image, such as brightening the clouds while keeping the sky a deep blue.

C. Maintaining a Natural Look

- Step 1: Avoid over-editing, which can make the clouds look unnatural. Aim for a balance that enhances the photo while retaining its natural beauty.

- Step 2: Compare the edited image with the original to ensure the changes are subtle and realistic, preserving the integrity of the scene.

By mastering camera settings, employing effective composition techniques, and applying thoughtful editing, you can capture the beauty of clouds in stunning photographs. This comprehensive guide will help you document and share your cloud spotting experiences with the world, showcasing the captivating and ever-changing sky.

9.2 The Art of Sketching Clouds

Sketching clouds can be a meditative and rewarding way to capture their beauty. This section provides comprehensive, step-by-step guidance on the materials needed, basic techniques, and advanced tips for creating stunning cloud sketches.

9.2.1 Materials Needed

Having the right materials is essential for effective cloud sketching. Here's a detailed list of what you'll need:

A. Basic Supplies

- Sketching Paper: Choose high-quality, smooth sketching paper that can handle both light and heavy pencil strokes without smudging or tearing.

- Pencils: Use a range of pencils from H (hard) to B (soft) for varying shades and textures. Common choices include HB, 2B, 4B, and 6B.

- Erasers: Have both a regular eraser for corrections and a kneaded eraser for subtle adjustments and highlights.

- Blending Tools: Use tortillons or blending stumps for smooth transitions between light and dark areas.

B. Optional Tools

- Charcoal or Pastels: These can add a different texture and depth to your sketches. Charcoal provides rich, dark lines, while pastels offer vibrant colors.

- Fixative Spray: This can help set your finished sketch, preventing smudging and preserving your work.

9.2.2 Basic Techniques

Mastering basic sketching techniques is crucial for accurately depicting clouds. Follow these steps to get started:

A. Observing Clouds

- Step 1: Spend time observing clouds, noting their shapes, movements, and textures. Pay attention to how light and shadow play on their surfaces.

- Step 2: Take reference photos if needed, capturing specific cloud formations to study later.

B. Sketching Outlines

- Step 1: Start with light pencil strokes to outline the basic shape and structure of the clouds. Focus on the overall form rather than getting lost in details initially.

- Step 2: Draw softly and loosely, allowing your hand to move freely to capture the fluid and dynamic nature of clouds.

C. Adding Details

- Step 1: Gradually add details to your sketch, such as the intricate edges and varying densities within the clouds. Use different pencil grades to create depth and dimension.

- Step 2: Emphasize the contrasts between light and dark areas to bring out the cloud's texture and volume.

D. Blending and Texturing

- Step 1: Use blending tools to smooth transitions between light and dark areas, creating a realistic, fluffy appearance. Blend lightly to avoid over-smoothing.

- Step 2: Add texture by varying your pencil strokes. Experiment with techniques like cross-hatching, stippling, and circular motions to achieve different effects.

9.2.3 Advanced Tips

Once you're comfortable with basic techniques, these advanced tips will help you refine your cloud sketches:

A. Light and Shadow

- Step 1: Study how light affects clouds at different times of the day. Practice sketching clouds in morning, afternoon, and evening light to capture the variations in color and shadow.

- Step 2: Use a soft pencil for shadows and a hard pencil for highlights, creating a realistic interplay of light and dark.

B. Atmospheric Perspective

- Step 1: Create a sense of depth by using lighter tones for distant clouds and darker, more detailed strokes for closer clouds. This technique helps convey the vastness of the sky.

- Step 2: Incorporate elements of the landscape, such as mountains or trees, to provide context and scale to your cloud sketches.

C. Experimentation

- Step 1: Experiment with different mediums, such as watercolor or ink, to add variety and interest to your cloud sketches. Watercolor can create soft, diffuse effects, while ink can add sharp contrast.

- Step 2: Try different techniques, like wet-on-wet watercolor for soft edges or ink washes for dramatic contrasts. These methods can help you discover new ways to depict clouds.

D. Continuous Learning

- Step 1: Study the works of other artists who specialize in cloud sketching or landscape art. Analyze their techniques and try to incorporate some of their methods into your own work.

- Step 2: Practice regularly to improve your skills. Set aside dedicated time for sketching and experiment with different styles and approaches.

By following these detailed, structured, and practical steps, you can master the art of sketching clouds. This guide will help you develop your skills, allowing you to capture the ephemeral beauty of clouds on paper and enhance your overall cloud spotting experience.

9.3 Sharing Your Cloud Spotting Experiences Online

Sharing your cloud spotting experiences online can be an enriching part of your journey, allowing you to connect with others, inspire fellow enthusiasts, and document your adventures. This section provides a comprehensive guide on using social media platforms, blogging, and building an online community to share your cloud spotting experiences effectively.

9.3.1 Social Media Platforms

Social media platforms offer powerful tools to share your cloud spotting experiences with a wide audience. Here's how to make the most of them:

4. Choosing the Right Platforms

- Step 1: Instagram: Ideal for sharing high-quality photos and short videos. Use hashtags like #cloudspotting, #clouds, and #skywatching to reach a broader audience.

- Step 2: TikTok: Great for creating engaging short videos and reaching a younger audience. Share time-lapse videos of cloud formations or quick tips on cloud spotting.

- Step 3: Facebook: Useful for longer posts and engaging with community groups. Join or create cloud spotting groups to connect with like-minded individuals.

- Step 4: Twitter: Perfect for sharing quick updates, weather observations, and connecting with the meteorology community.

B. Engaging Content

- Step 1: Post regularly with high-quality photos and videos of your cloud spotting experiences. Aim for consistency to keep your audience engaged.

- Step 2: Use engaging captions that provide context or interesting facts about the clouds you're sharing. Ask questions to encourage interaction.

- Step 3: Utilize stories and live sessions to share real-time cloud spotting adventures. This provides an immediate and personal connection with your audience.

C. Interacting with Followers

- Step 1: Respond to comments and messages promptly to build a connection with your followers. Engagement is key to growing a loyal audience.

- Step 2: Collaborate with other cloud spotters and meteorology enthusiasts. Cross-promote each other's content to expand your reach.

- Step 3: Participate in or create hashtag challenges related to cloud spotting to increase visibility and foster community interaction.

9.3.2 Blogging Your Journey

Blogging is a powerful way to document and share your cloud spotting journey in a more detailed and personal manner.

A. Starting a Blog

- Step 1: Choose a blogging platform like WordPress, Blogger, or Squarespace. These platforms offer user-friendly interfaces and customizable templates.

- Step 2: Create an appealing blog design that reflects your passion for cloud spotting. Use a clean, visually appealing layout with easy navigation.

B. Writing Engaging Posts

- Step 1: Share detailed accounts of your cloud spotting adventures, including locations, weather conditions, and personal reflections. This adds depth to your posts and makes them more relatable.

- Step 2: Include high-quality photos and sketches to illustrate your posts. Visuals enhance the reading experience and provide context to your observations.

- Step 3: Write how-to guides, such as tips for beginners, equipment recommendations, or step-by-step instructions for identifying different cloud types.

C. Building Readership

- Step 1: Promote your blog on social media and cloud spotting forums. Share links to your posts and encourage your followers to visit your blog.

- Step 2: Engage with your readers by responding to comments and encouraging feedback. This interaction builds a sense of community and loyalty.

- Step 3: Collaborate with other bloggers or influencers in the meteorology and nature niches. Guest posts and joint projects can help you reach a wider audience.

9.3.3 Building an Online Community

Building an online community allows you to connect with fellow cloud spotters, share knowledge, and foster a supportive environment.

A. Creating a Community Space

- Step 1: Use platforms like Facebook Groups, Discord, or Reddit to create a dedicated space for cloud spotting enthusiasts. Ensure the group is easy to find and join.

- Step 2: Set clear guidelines and objectives for the community to foster a respectful and focused environment. Encourage members to share their experiences, photos, and questions.

B. Hosting Online Events

- Step 1: Organize virtual events such as live cloud spotting sessions, webinars, or Q&A sessions with meteorologists. Use tools like Zoom or Facebook Live for these events.

- Step 2: Promote these events on your social media and blog to attract participants. Send reminders and provide clear instructions on how to join.

C. Encouraging Collaboration

- Step 1: Foster a collaborative environment where members can work on projects together, such as creating a cloud spotting guide or organizing group outings.

- Step 2: Highlight and celebrate the achievements of community members to keep the group motivated and engaged. Share member contributions, such as photos or articles, and acknowledge their efforts.

By effectively using social media platforms, blogging, and building an online community, you can share your cloud spotting experiences with a broader audience and inspire others to appreciate the beauty of the sky. This comprehensive guide will help you connect with fellow enthusiasts, document your journey, and create a supportive network of cloud spotters.

Chapter 9 Review

Chapter 9: Capturing the Beauty of Clouds provides a comprehensive guide on how to document and share your cloud spotting experiences through photography, sketching, and online sharing. This chapter offers practical, step-by-step advice on capturing stunning images and sketches of clouds and effectively sharing your passion with a wider audience.

9.1 Photography Tips and Tricks

Mastering cloud photography involves understanding your camera settings, employing effective composition techniques, and applying thoughtful editing.

- *Camera Settings:*

  - ISO: Set to 100-200 to minimize noise.

  - Aperture: Use a small aperture (f/8 to f/16) for deep depth of field.

  - Shutter Speed: Adjust based on lighting; faster speeds (1/250s or higher) for sharp cloud details.

  - White Balance: Set to daylight or adjust manually for accurate colors.

  - Filters: Use polarizing filters for reducing glare and enhancing contrast, and graduated neutral density (ND) filters for balanced exposure.

- *Composition Techniques:*

  - Rule of Thirds: Place key elements along the grid lines or intersections.

  - Leading Lines: Use natural lines to guide the viewer's eye.

  - Foreground Interest: Include elements like trees or mountains for context and scale.

- Framing: Use natural frames to highlight clouds without obstructing the view.

- Editing Tips:

- Basic Adjustments: Use software like Lightroom or Photoshop to adjust exposure, contrast, and saturation.

- Advanced Techniques: Enhance clarity and definition with clarity and dehaze sliders, and apply selective editing for emphasis.

- Maintaining Natural Look: Avoid over-editing to keep clouds looking realistic.

9.2 The Art of Sketching Clouds

Sketching clouds requires the right materials, mastering basic techniques, and advancing with practice and experimentation.

- Materials Needed:

- Basic Supplies: High-quality sketching paper, a range of pencils (HB to 6B), regular and kneaded erasers, and blending tools like tortillons.

- Optional Tools: Charcoal, pastels, and fixative spray for different textures and preservation.

- Basic Techniques:

- Observing Clouds: Spend time noting shapes, movements, and light/shadow interactions.

- Sketching Outlines: Start with light strokes to outline basic shapes and structure.

- Adding Details: Gradually add details, emphasizing contrasts and textures.

- Blending and Texturing: Use blending tools for smooth transitions and vary pencil strokes for different effects.

- *Advanced Tips:*

  - Light and Shadow: Study and practice capturing clouds at different times of the day.

  - Atmospheric Perspective: Use lighter tones for distant clouds and darker, detailed strokes for closer clouds.

  - Experimentation: Try different mediums and techniques like watercolor or ink.

  - Continuous Learning: Study other artists' works and practice regularly.

9.3 Sharing Your Cloud Spotting Experiences Online

Sharing your cloud spotting experiences online involves using social media, blogging, and building an online community.

- *Social Media Platforms:*

  - Choosing the Right Platforms: Instagram for photos, TikTok for videos, Facebook for community groups, and Twitter for quick updates.

  - Engaging Content: Post regularly with high-quality visuals and engaging captions, use stories and live sessions.

  - Interacting with Followers: Respond to comments, collaborate with others, and participate in hashtag challenges.

- *Blogging Your Journey:*

- Starting a Blog: Use platforms like WordPress or Blogger, and create an appealing design.

- Writing Engaging Posts: Share detailed accounts of your experiences, include high-quality visuals, and write how-to guides.

- Building Readership: Promote your blog on social media, engage with readers, and collaborate with other bloggers.

- Building an Online Community:

- Creating a Community Space: Use platforms like Facebook Groups or Discord, and set clear guidelines.

- Hosting Online Events: Organize live sessions, webinars, or Q&A sessions, and promote them effectively.

- Encouraging Collaboration: Foster a collaborative environment and celebrate community achievements.

By mastering the techniques for capturing clouds through photography and sketching and effectively sharing your experiences online, you can inspire others and build a community of cloud spotting enthusiasts. This chapter equips you with the tools and knowledge to document and share the beauty of clouds, making your cloud spotting journey even more fulfilling.

Part 5: The Artistic and Cultural Side of Clouds

Chapter 10: Clouds in Art and Media

Clouds have been a profound source of inspiration in the realm of art and media, shaping the creative expression of artists, photographers, and cultural movements throughout history. This chapter delves into the influential role of clouds in famous paintings, the work of renowned photographers, and their impact on various artistic movements.

10.1 Famous Paintings and Photographs

10.1.1 Iconic Paintings

Clouds have played a crucial role in many iconic paintings, adding depth, emotion, and drama to the artwork.

A. Renaissance to Romanticism

1. Leonardo da Vinci:

- Observation and Detail: Leonardo's meticulous observations of nature extended to his cloud depictions, integrating them seamlessly into the background of his landscapes, enhancing the atmospheric perspective.

- Study: Examine "The Annunciation," where Leonardo's clouds contribute to the serene yet profound background, emphasizing divine presence.

2. J.M.W. Turner:

- Emotion and Sublime: Turner used clouds to convey powerful emotions and the sublime beauty of nature. His dramatic skies often reflect turbulent weather, symbolizing human emotions and nature's power.

- Study: Look at "Rain, Steam, and Speed – The Great Western Railway," where the clouds blend with steam, creating a dynamic, atmospheric effect.

B. Impressionism

1. Claude Monet:

- Light and Color: Monet's Impressionist works capture the transient effects of light and atmosphere, with clouds playing a vital role in setting the scene's mood.

- Study: Analyze "Impression, Sunrise," where the clouds merge with the mist and sunlight, creating a hazy, ethereal quality.

2. Vincent van Gogh:

- Expressive Brushstrokes: Van Gogh's use of bold, swirling brushstrokes brings clouds to life, expressing movement and energy.

- Study: Observe "Starry Night," where the turbulent sky with swirling clouds adds to the painting's emotional intensity.

C. Modern and Contemporary Art

1. Georgia O'Keeffe:

- Abstract Forms: O'Keeffe depicted clouds with bold, abstract forms, reflecting their changing nature and emphasizing simplicity and essence.

- Study: Look at "Sky Above Clouds IV," where she abstracts clouds into rhythmic, layered shapes that stretch across the canvas.

2. *Contemporary Artists:*

   - Mixed Media and Digital Art: Modern artists use various media, including digital art, to create innovative representations of clouds, pushing the boundaries of traditional art.

   - Study: Explore works by contemporary artists who integrate technology, such as digital painting and augmented reality, to depict clouds in new, imaginative ways.

10.1.2 Renowned Photographers

Photographers have captured the beauty of clouds through different techniques and styles, highlighting their ever-changing nature.

A. Early Pioneers

1. Ansel Adams:

   - Black and White Mastery: Adams' black-and-white photographs emphasize the textures and contrasts of clouds, capturing their grandeur.

   - Study: Examine "Clearing Winter Storm," where the dramatic clouds enhance the majestic landscape of Yosemite.

2. Alfred Stieglitz:

   - Equivalents Series: Stieglitz's "Equivalents" series focuses on the abstract forms of clouds, expressing emotions through their shapes.

   - Study: Analyze how Stieglitz uses clouds to evoke feelings and moods, treating them as subjects in their own right.

B. Modern Masters

1. Hiroshi Sugimoto:

- Minimalism and Abstraction: Sugimoto's minimalist approach captures the ethereal quality of clouds, often blurring the line between reality and abstraction.

- Study: Look at his "Seascapes" series, where the horizon line divides the sea and sky, emphasizing the clouds' soft, dreamy forms.

2. Michael Kenna:

- Long Exposure: Kenna's long-exposure photographs create smooth, serene cloudscapes, often with a surreal, otherworldly feel.

- Study: Explore his works where clouds are captured with extended shutter speeds, creating a sense of tranquility and movement.

C. Techniques and Styles

1. Long Exposure:

- Technique: Use long exposure to capture the movement of clouds, creating smooth, flowing images that highlight their dynamic nature.

- Application: Experiment with shutter speeds and neutral density filters to achieve the desired effect.

2. Time-Lapse Photography:

- Technique: Capture a series of images over time to create time-lapse videos, showcasing the evolution and movement of clouds.

- Application: Set up your camera on a stable surface, use interval shooting, and compile the images into a video using editing software.

3. Infrared Photography:

- Technique: Use infrared filters to capture clouds in a unique way, often rendering them with striking contrast and ethereal quality.

- Application: Experiment with different infrared filters and post-processing techniques to bring out the unique characteristics of clouds.

10.1.3 Artistic Movements

Different artistic movements have interpreted and depicted clouds in various ways, reflecting cultural and historical contexts.

A. Romanticism

1. Emotional Intensity:

- Focus: Romantic artists used clouds to evoke powerful emotions and the sublime beauty of nature, often portraying dramatic and turbulent skies.

- Study: Examine how artists like J.M.W. Turner and Caspar David Friedrich used clouds to convey feelings of awe, melancholy, and transcendence.

2. Symbolism:

- Focus: Clouds often symbolized the uncontrollable forces of nature and the human spirit's connection to the natural world.

- Study: Analyze the symbolic meanings behind the cloud-filled skies in Romantic paintings, reflecting the artists' fascination with nature's power and mystery.

B. Impressionism

1. Light and Atmosphere:

- Focus: Impressionist artists captured the fleeting effects of light and atmosphere, with clouds playing a crucial role in setting the scene's mood.

- Study: Look at how artists like Claude Monet and Vincent van Gogh used clouds to explore the interplay of light and color.

2. Brushstrokes and Color:

- Focus: Impressionists used loose, dynamic brushstrokes and vibrant colors to depict clouds, emphasizing their transient and ephemeral nature.

- Study: Observe the techniques used to create a sense of movement and immediacy in cloud paintings.

C. Surrealism

1. Dreamlike Imagery:

- Focus: Surrealist artists like René Magritte used clouds in dreamlike, fantastical compositions to challenge perceptions of reality.

- Study: Explore how Surrealists used clouds to create a sense of wonder, mystery, and the uncanny.

2. Psychological Interpretation:

- Focus: Clouds in Surrealist art often have symbolic meanings, representing subconscious thoughts and emotions.

- Study: Analyze the psychological and symbolic interpretations of clouds in Surrealist works, reflecting the artists' exploration of the inner mind.

By understanding the role of clouds in art and media, you can appreciate their cultural significance and draw inspiration for your own creative endeavors. This chapter provides a detailed, structured guide to exploring how clouds have influenced and enriched various artistic and cultural expressions.

10.2 Clouds in Literature and Film

Clouds have been a powerful motif in literature and film, symbolizing a range of emotions, themes, and cultural ideas. This section explores how clouds have been depicted in various literary works and films, and delves into their symbolic meanings.

10.2.1 Literary Works

Clouds have been used by writers throughout history to symbolize emotions, set scenes, and convey deeper meanings.

A. Classic Literature

1. William Wordsworth

   - Works: Wordsworth often used clouds in his poetry to reflect his deep connection with nature.

   - Example: In "I Wandered Lonely as a Cloud," clouds symbolize the poet's solitary yet reflective state.

   - Analysis: Explore how Wordsworth's description of clouds enhances the themes of solitude and the beauty of nature.

2. Emily Brontë

- Works: Brontë used clouds to set the dramatic and moody atmosphere in her novel "Wuthering Heights."

- Example: The changing weather and cloud formations mirror the tumultuous emotions of the characters.

- Analysis: Examine how the clouds contribute to the novel's gothic ambiance and emotional landscape.

B. Modern and Contemporary Literature

1. Italo Calvino

- Works: In "Clouds" from "Cosmicomics," Calvino uses clouds to explore abstract and philosophical ideas.

- Example: Clouds are personified and become a medium through which the protagonist experiences and understands the world.

- Analysis: Discuss how Calvino's use of clouds reflects on the nature of perception and existence.

2. Haruki Murakami

- Works: Murakami often incorporates clouds in his novels to create surreal and dreamlike atmospheres.

- Example: In "Kafka on the Shore," clouds are part of the natural world that interacts with the characters' inner lives.

- Analysis: Analyze how clouds enhance the surreal and introspective elements of Murakami's storytelling.

10.2.2 Filmography

In film, clouds can be visually stunning elements that enhance the narrative, evoke emotions, and symbolize themes.

A. Iconic Films

1. Terrence Malick's "The Tree of Life"

   - Usage: Malick uses clouds extensively to symbolize themes of creation, existence, and the passage of time.

   - Example: The film opens with cosmic imagery, including clouds, to set a contemplative tone.

   - Analysis: Examine how the imagery of clouds contributes to the film's exploration of life's profound questions.

2. Hayao Miyazaki's Animated Films

   - Usage: Miyazaki often features clouds prominently to create whimsical, magical atmospheres.

   - Example: In "Castle in the Sky," clouds play a crucial role in the visual storytelling, enhancing the sense of adventure and wonder.

   - Analysis: Discuss how Miyazaki's depiction of clouds adds depth to his fantastical worlds and narratives.

B. Visual Techniques

1. Time-Lapse and CGI

   - Technique: Filmmakers use time-lapse photography to capture the movement of clouds, adding a dynamic and dramatic effect to scenes.

- Application: Understand how time-lapse can convey the passage of time or create a sense of urgency.

  - Example: Analyze the use of time-lapse cloud footage in "Koyaanisqatsi" to evoke themes of nature and human impact.

2. Color Grading and Lighting

  - Technique: The color and lighting of clouds can dramatically alter the mood of a scene.

  - Application: Learn how filmmakers adjust these elements to match the emotional tone of the narrative.

  - Example: Study the stormy clouds in "Twister" and how they heighten the tension and danger of the scenes.

10.2.3 Symbolism in Media

Clouds are rich with symbolic meanings in various forms of media, often reflecting broader themes and cultural concepts.

A. Symbolic Meanings

1. Freedom and Transcendence

  - Focus: Clouds often symbolize freedom, dreams, and transcendence due to their high, unattainable nature.

  - Example: In "The Shawshank Redemption," clouds in the final scene symbolize the protagonist's newfound freedom and hope.

  - Analysis: Explore how clouds are used to visually represent themes of escape and aspiration.

2. Change and Impermanence

- Focus: The ever-changing nature of clouds makes them apt symbols for transformation and the fleeting nature of life.

- Example: In "American Beauty," the cloud formations during the film's conclusion symbolize the protagonist's acceptance of life's transience.

- Analysis: Discuss how the transient qualities of clouds reflect characters' personal journeys and changes.

B. Emotional and Psychological Impact

1. Mood and Atmosphere

- Focus: Clouds can significantly influence the mood and atmosphere of a scene, evoking emotions ranging from peace to foreboding.

- Example: In "Blade Runner," the dark, cloud-filled sky creates a dystopian and oppressive atmosphere.

- Analysis: Examine how cloud imagery is used to amplify the emotional tone of films.

2. Inner States and Symbolic Reflection

- Focus: Clouds can mirror the inner states of characters, symbolizing their emotions and mental states.

- Example: In "The Lovely Bones," the protagonist's visions of the afterlife are filled with surreal cloudscapes that reflect her emotional journey.

- Analysis: Analyze how clouds serve as a visual metaphor for the characters' internal experiences.

By understanding how clouds are used in literature and film, you can appreciate their symbolic depth and artistic significance. This chapter provides a structured guide to exploring the cultural impact of clouds, enhancing your cloud spotting journey with a richer context.

10.3 How Clouds Inspire Creativity

10.3.1 Artistic Expression

Step 1: Observation and Sketching

Find a comfortable place with a clear view of the sky, such as a park or an open field. Bring a sketchbook, pencils, and any other preferred art supplies. Look at the clouds and start sketching their shapes and forms, focusing on capturing their essence rather than exact details. Experiment with different artistic mediums like watercolors, pastels, or digital tools to capture the fluid nature of clouds.

Step 2: Interpretation and Abstraction

Spend time identifying shapes within the clouds, such as animals, faces, or abstract forms. Use these shapes as the basis for more abstract pieces, letting your imagination transform them into unique artworks. Allow the mood of the sky to influence your work. Stormy clouds might inspire dramatic art, while fluffy clouds on a sunny day could lead to whimsical pieces.

Step 3: Using Photography

Use a camera or smartphone with high resolution, experimenting with exposure and focus settings. Apply composition techniques such as the rule of thirds or leading lines to create compelling cloud photographs. Use photo editing software to enhance your images, adjusting brightness, contrast, and saturation to highlight cloud formations.

10.3.2 Personal Creativity

Step 1: Journaling and Reflection

Keep a cloud journal to document your observations and reflections. Write detailed descriptions of the clouds, including shapes, movements, and the feelings they evoke. Make cloud journaling a daily habit to develop a deeper connection with the sky and find new sources of inspiration.

Step 2: Creative Writing

Use clouds as prompts for creative writing. Imagine stories based on the shapes and movements of the clouds you observe. Write poems inspired by clouds, focusing on their transient nature and the emotions they evoke. Employ clouds as metaphors and similes in your writing to represent various concepts, from freedom to mystery.

Step 3: Mindfulness and Meditation

Spend time mindfully observing clouds, focusing on their movement and how they change shape. Incorporate cloud observation into your meditation routine, letting your thoughts drift with the clouds to promote relaxation and mental clarity. Use clouds as a tool for creative visualization, imagining yourself interacting with the clouds and exploring their shapes and forms.

10.3.3 Collaborative Projects

Step 1: Community Art Projects

Organize community art projects where participants create cloud-inspired artworks together. These can include collaborative murals, group photography exhibitions, or communal sculpture projects. Host workshops focused on cloud observation and artistic creation, providing materials and guidance for participants. Display the collective works in local galleries, community centers, or public spaces to celebrate the creativity inspired by clouds.

Step 2: Group Creative Writing and Storytelling

Engage in group storytelling sessions using clouds as the starting point. Each participant can add to the story based on their interpretation of cloud shapes and forms. Create shared cloud journals where multiple people can document their observations and reflections. Use online platforms to collaborate on cloud-inspired writing projects, allowing participants to contribute from different locations and share their unique perspectives.

Step 3: Educational Programs

Develop educational programs for schools that incorporate cloud spotting into the curriculum. Encourage students to observe clouds and express their creativity through art and writing. Offer workshops for adults that focus on the creative potential of cloud spotting, including art classes, writing workshops, and mindfulness sessions. Partner with local organizations to bring cloud-inspired creative projects to underserved communities, fostering creativity and a sense of connection with nature.

By incorporating clouds into various forms of creative expression, individuals and groups can find endless inspiration and new ways to connect with the natural world. Whether through art, writing, or collaborative projects, cloud spotting can unlock a wealth of creative potential.

Chapter 10 Review

Clouds have significantly influenced art and media, inspiring numerous creative works.

10.1 Famous Paintings and Photographs

Iconic Paintings

- Leonardo da Vinci: Detailed clouds enhance perspective.

  - Example: "The Annunciation"

- J.M.W. Turner: Clouds convey emotions and nature's power.

  - Example: "Rain, Steam, and Speed"

- Claude Monet: Clouds capture transient light and mood.

  - Example: "Impression, Sunrise"

- Vincent van Gogh: Swirling clouds express movement and intensity.

  - Example: "Starry Night"

- Georgia O'Keeffe: Abstract cloud forms.

  - Example: "Sky Above Clouds IV"

Renowned Photographers

- Ansel Adams: Black-and-white photos emphasize cloud textures.

  - Example: "Clearing Winter Storm"

- Alfred Stieglitz: Emotional cloud forms in "Equivalents" series.

- Hiroshi Sugimoto: Minimalist, abstract cloud photography.

  - Example: "Seascapes"

- Michael Kenna: Long-exposure photos create serene cloudscapes.

10.2 Clouds in Literature and Film

Literary Works

- William Wordsworth: Clouds symbolize solitude and nature.

  - Example: "I Wandered Lonely as a Cloud"

- Emily Brontë: Clouds set dramatic, moody scenes.

  - Example: "Wuthering Heights"

- Italo Calvino: Clouds explore perception and existence.

  - Example: "Clouds" from "Cosmicomics"

- Haruki Murakami: Clouds enhance surreal storytelling.

  - Example: "Kafka on the Shore"

Filmography

- Terrence Malick: Clouds symbolize creation and time.

  - Example: "The Tree of Life"

- Hayao Miyazaki: Clouds create magical atmospheres.

  - Example: "Castle in the Sky"

10.3 How Clouds Inspire Creativity

Artistic Expression

- Observation and Sketching: Sketch cloud shapes.

- Interpretation and Abstraction: Use cloud shapes for abstract art.

- Photography: Capture and enhance cloud images.

Personal Creativity

- Journaling and Reflection: Document cloud observations.

- Creative Writing: Use clouds as prompts for stories and poems.

- Mindfulness and Meditation: Observe clouds for relaxation.

Collaborative Projects

- Community Art Projects: Create cloud-inspired art together.

- Group Creative Writing: Collaborate on cloud-inspired stories.

- Educational Programs: Incorporate cloud spotting into curriculums.

Clouds offer endless inspiration for art, literature, and film, encouraging creativity and cultural expression.

Chapter 11: Creating Cloud-Inspired Art

Clouds have always captivated the imagination with their ever-changing forms and ephemeral beauty. For centuries, artists have drawn inspiration from the sky, translating the majesty of clouds into various forms of art. In this chapter, we explore how you can harness the inspiration from cloud spotting to create your own cloud-inspired art. From painting and sculpting to digital creations, we'll guide you through several DIY projects. Additionally, we'll discuss how to share your art with others through cloud-themed art shows and collaborations with fellow cloud enthusiasts. Whether you're a seasoned artist or a beginner, this chapter will provide you with the tools and inspiration to bring the beauty of clouds into your creative endeavors.

11.1 DIY Cloud Art Projects

Creating art inspired by clouds can be a rewarding and therapeutic experience. Whether you are painting, sculpting, or using digital tools, translating the beauty of clouds into your artwork can enhance your appreciation for the natural world and boost your creativity. This section will guide you through three types of cloud-inspired art projects: painting clouds, sculpting clouds, and creating digital cloud art.

11.1.1 Painting Clouds

Step 1: Gathering Materials

- Choose your painting medium: watercolor, acrylic, or oil paints.

- Get quality brushes of various sizes, a palette, and a painting surface such as canvas, paper, or wood.

Step 2: Observation

- Spend time observing clouds in different weather conditions and times of day.

Take photographs or make quick sketches of clouds to use as references for your painting.

Step 3: Initial Sketch

Lightly sketch the basic shapes of the clouds on your painting surface with a pencil.

Focus on capturing the overall form and flow of the clouds rather than detailed features.

Step 4: Laying Down the Base Colors

Start with the sky, using a gradient from light to dark to create depth.

Apply a base color for the clouds, typically a mix of white with hints of gray or blue.

Step 5: Adding Details

- Build up layers of paint to add depth and dimension to your clouds.

- Use a variety of brush strokes to mimic the texture and movement of clouds.

- Blend the edges of the clouds into the sky to create a natural, airy effect.

Step 6: Final Touches

- Add highlights and shadows to the clouds to enhance their three-dimensional appearance.

- Step back frequently to view your work from a distance and make adjustments as needed.

- Consider adding subtle color variations to reflect the light and weather conditions.

11.1.2 Sculpting Clouds

Step 1: Choosing Materials

- Select materials such as clay, papier-mâché, or lightweight foam.

- Gather sculpting tools appropriate for your chosen material, such as modeling tools, knives, and smoothing tools.

Step 2: Planning Your Sculpture

- Decide on the size and shape of your cloud sculpture.

- Make a rough sketch or model to plan the overall design.

Step 3: Building the Structure

- For larger sculptures, create an armature or framework using wire or cardboard to support the structure.

- If using papier-mâché, create a base form with crumpled paper or balloons before applying layers of papier-mâché.

Step 4: Adding Texture and Details

- Use sculpting tools to add texture and details to the surface of your cloud.

- For clay, use tools to carve and shape the surface, creating a fluffy, dynamic appearance.

For papier-mâché or foam, use additional layers and shaping techniques to build up texture.

Step 5: Painting and Finishing

- Once the sculpture is dry and stable, paint it with acrylic paints to add color and depth.

- Use a variety of whites, grays, and blues to create a realistic cloud effect.

- Apply a protective sealant if desired to preserve the sculpture.

11.1.3 Digital Cloud Art

Step 1: Choosing Software

- Select digital art software that suits your skill level and style, such as Adobe Photoshop, Corel Painter, or Procreate.

- Ensure you have a digital tablet and stylus for more precise control.

Step 2: Setting Up Your Workspace

- Open your chosen software and set up a new canvas.

- Adjust the canvas size and resolution to your preference.

Step 3: Creating the Base Layer

- Start with a background layer, creating a gradient sky using soft brushes.

- Add a new layer for the clouds and choose a soft round brush with varying opacity settings.

Step 4: Sketching Clouds

- Use a light gray or white color to sketch the basic shapes of the clouds.

- Focus on creating smooth, flowing lines to capture the natural movement of clouds.

Step 5: Adding Depth and Detail

- Use different brushes and opacity settings to add layers to the clouds.

- Experiment with blending modes and layer effects to create depth and dimension.

- Incorporate highlights and shadows to enhance the three-dimensional look of the clouds.

Step 6: Refining and Finalizing

- Refine the edges of the clouds using erasers and blending tools to create a natural look.

- Add final details such as color variations and soft glow effects to make the clouds more realistic.

- Save your work in high resolution and consider printing it for physical display.

By following these detailed steps, you can create stunning cloud-inspired art that captures the beauty and essence of clouds, whether through traditional painting, sculpting, or digital techniques.

11.2 Hosting a Cloud-Themed Art Show

Hosting a cloud-themed art show is a wonderful way to share your passion for clouds and creativity with a broader audience. This section provides a comprehensive, step-by-step guide to planning, organizing, and executing a successful cloud-themed art show.

11.2.1 Planning and Organization

Step 1: Define Your Vision and Goals

- Determine the purpose of your art show. Is it to showcase personal work, feature multiple artists, raise awareness about cloud spotting, or a combination of these?

- Establish clear goals, such as the number of attendees, amount of art to be displayed, and any fundraising or community engagement objectives.

Step 2: Set a Budget

- Create a detailed budget covering all aspects of the event, including venue rental, marketing, materials, refreshments, and any additional services.

- Seek out sponsors or partners who may be interested in supporting the event financially or through in-kind contributions.

Step 3: Choose a Venue

- Select a venue that fits your vision and budget. Options include galleries, community centers, schools, parks, or even outdoor spaces under clear skies.

- Ensure the venue has adequate space for displaying art, accommodating guests, and hosting any additional activities.

Step 4: Set a Date and Time

- Pick a date that allows enough time for planning and preparation. Consider any local events or holidays that may conflict with your chosen date.

- Decide on the duration of the show—whether it will be a one-day event or span several days or weeks.

Step 5: Curate the Art

- Decide whether the show will feature only your work or include pieces from other artists. If including others, create a call for submissions with clear guidelines.

- Select a diverse range of artworks that highlight various interpretations and styles of cloud-inspired art.

Step 6: Organize Logistics

- Coordinate with artists to collect and transport their work to the venue.

- Arrange for any necessary equipment, such as easels, display cases, lighting, and sound systems.

- Plan for setup and takedown, assigning specific tasks to volunteers or staff.

11.2.2 Displaying Your Work

Step 1: Plan the Layout

- Sketch a floor plan of the venue to determine the best arrangement for displaying the artwork.

- Consider the flow of foot traffic, ensuring there is enough space for guests to move comfortably and view each piece.

Step 2: Prepare the Artwork

- Ensure all artworks are ready for display, with frames, stands, or mounts as needed.

- Include labels or placards for each piece, providing information about the artist, title, medium, and a brief description or inspiration behind the work.

Step 3: Arrange the Display

- Start with the focal points, placing the most impactful or central pieces in prominent positions.

- Group artworks thematically or by artist to create a cohesive viewing experience.

- Use a variety of display methods, such as hanging art on walls, placing sculptures on pedestals, and setting up digital displays for digital art.

Step 4: Enhance the Ambiance

- Use lighting strategically to highlight key pieces and create an inviting atmosphere.

- Consider background music that complements the theme without overpowering the conversation.

- Add decorative elements that enhance the cloud theme, such as fabric drapes, cloud cutouts, or thematic centerpieces.

11.2.3 Engaging Your Audience

Step 1: Promote the Event

- Utilize various marketing channels to spread the word about your art show. This can include social media, email newsletters, local media, community bulletin boards, and flyers.

- Create an event page or website with all the details, including the date, time, location, and a preview of the artwork.

Step 2: Prepare Opening Remarks

- Write and practice a brief speech to welcome guests, introduce the theme of the show, and provide context for the artwork on display.

- Acknowledge any collaborators, sponsors, and participating artists.

Step 3: Plan Interactive Elements

- Incorporate activities that engage guests, such as live demonstrations, artist talks, or guided tours of the exhibition.

- Set up a guest book or interactive wall where attendees can leave comments, reflections, or doodles inspired by the art.

Step 4: Offer Refreshments

- Provide light refreshments and beverages to create a welcoming and sociable atmosphere. Consider thematic treats that tie into the cloud theme.

- Arrange for tables and seating areas where guests can relax and discuss the art.

Step 5: Foster Networking and Connection

- Encourage artists and guests to mingle and share their experiences and inspirations.

- Provide opportunities for networking, such as designated mingling areas or informal meet-and-greet sessions.

Step 6: Collect Feedback and Follow Up

- Have feedback forms available for guests to share their thoughts and suggestions.

- Follow up with attendees and artists after the event, thanking them for their participation and support. Share photos and highlights from the event to keep the momentum going.

By following these steps, you can host a successful cloud-themed art show that not only showcases beautiful artwork but also fosters a sense of community and appreciation for the beauty of clouds.

11.3 Collaborating with Other Cloud Spotters

Collaborating with other cloud spotters can greatly enhance your cloud spotting experience. It provides opportunities to share knowledge, gain new perspectives, and create collaborative art projects. This section will guide you through organizing group projects, engaging in community art initiatives, and leveraging online platforms for collaboration.

11.3.1 Group Projects

Step 1: Identify Your Group

- Reach out to local cloud spotting clubs, art groups, or online communities to find like-minded individuals.

- Create a list of interested participants, including their contact information and areas of interest.

Step 2: Plan the Project

- Decide on the type of group project you want to undertake, such as a collaborative art piece, a group photo exhibition, or a cloud spotting field trip.

- Set clear goals and objectives for the project, including timelines and expected outcomes.

Step 3: Assign Roles and Responsibilities

- Divide tasks among group members based on their skills and interests.

- Assign roles such as project coordinator, communications manager, and activity leader to ensure smooth execution.

Step 4: Schedule Regular Meetings

- Organize regular meetings to discuss progress, share ideas, and address any challenges.

- Use these meetings to build camaraderie and maintain momentum.

Step 5: Execute the Project

- Follow the planned steps and ensure everyone is contributing as agreed.

- Maintain open communication and flexibility to adapt to any changes or challenges that arise.

Step 6: Showcase the Results

- Plan an event or platform to display the results of your group project, such as an exhibition, presentation, or online gallery.

- Celebrate the collaboration and acknowledge everyone's contributions.

11.3.2 Community Art

Step 1: Define the Community Art Project

- Decide on the scope and purpose of the community art project, such as a mural, public art installation, or community art day.

- Identify the target audience and potential participants within the community.

Step 2: Secure Support and Resources

- Seek support from local businesses, art organizations, and community leaders.

- Gather necessary resources, such as funding, materials, and permissions for public space use.

Step 3: Promote the Project

Use local media, social media, and community bulletin boards to spread the word.

- Create flyers, posters, and an online event page to attract participants.

Step 4: Organize Workshops and Planning Sessions

- Host workshops to brainstorm ideas and plan the details of the project.

- Encourage community members to share their visions and artistic talents.

Step 5: Execute the Community Art Project

- Schedule dates for the actual creation of the art project.

- Ensure there are enough volunteers and coordinators to assist participants.

Step 6: Celebrate and Display

- Organize an unveiling event or exhibition to showcase the completed project.

- Invite the community to celebrate and engage with the artwork.

11.3.3 Online Collaborations

Step 1: Find Online Communities

- Join online platforms and forums dedicated to cloud spotting and art, such as social media groups, specialized forums, and creative collaboration websites.

- Participate actively to connect with other enthusiasts and find potential collaborators.

Step 2: Propose Collaborative Projects

- Share your ideas for collaborative projects and invite others to join.

- Be clear about the project's goals, timelines, and what kind of contributions you're seeking.

Step 3: Use Collaboration Tools

- Utilize online tools for collaboration, such as cloud storage for sharing files, project management apps for organizing tasks, and communication platforms for discussions.

- Ensure all participants have access to the necessary tools and understand how to use them.

Step 4: Maintain Communication

- Schedule regular virtual meetings or check-ins to discuss progress and share updates.

- Use email, messaging apps, and video calls to stay connected and keep everyone informed.

Step 5: Compile and Finalize the Project

- Collect and compile contributions from all participants.

- Use digital tools to edit, refine, and finalize the project, whether it's a digital art piece, an e-book, or an online gallery.

Step 6: Share and Promote

- Publish the final project on relevant online platforms, such as websites, social media, and online galleries.

- Promote the project through your networks and encourage participants to share it within their communities.

By following these detailed steps, you can successfully collaborate with other cloud spotters on a variety of creative projects. Whether working in person or online, these collaborations can lead to enriching experiences and beautiful, cloud-inspired art that can be shared and enjoyed by many.

Chapter 11 Review

Chapter 11 explores how cloud spotting can serve as a powerful source of inspiration for creative endeavors. This chapter is structured into three comprehensive sections, each offering detailed, practical guidance on turning your observations of clouds into artistic and collaborative projects.

11.1 DIY Cloud Art Projects

In this section, we provide a step-by-step guide on how to create cloud-inspired art using various mediums. Whether you are a seasoned artist or a beginner, thes projects are designed to help you capture the beauty and essence of clouds.

- Painting Clouds: Learn how to observe, sketch, and paint clouds using differen techniques and mediums, such as watercolor, acrylic, and oil paints. This part includes tips on choosing the right materials, creating depth and texture, and adding final touches to your artwork.

- Sculpting Clouds: Explore the world of three-dimensional art by sculpting clouds using materials like clay, papier-mâché, or lightweight foam. Detailed instructions cover the entire process, from planning and building the structure to adding texture, painting, and finishing your sculpture.

- Digital Cloud Art: Dive into the realm of digital art with guidance on creating cloud-inspired pieces using software like Adobe Photoshop, Corel Painter, or Procreate. This part provides practical tips on setting up your digital workspace, using various tools and brushes, and refining your digital artwork.

11.2 Hosting a Cloud-Themed Art Show

This section focuses on how to share your cloud-inspired art with a wider audience by hosting a cloud-themed art show. It covers every aspect of planning and executing a successful event.

- Planning and Organization: Learn how to define your vision and goals, set a budget, choose a venue, and schedule the event. This part also includes advice on curating the art, organizing logistics, and ensuring a smooth setup and takedown.

- Displaying Your Work: Discover how to effectively plan the layout of your art show, prepare and arrange the artwork, and enhance the ambiance of the venue. This part provides practical tips on lighting, background music, and decorative elements to create an inviting atmosphere.

- Engaging Your Audience: Gain insights into promoting your event, preparing opening remarks, and planning interactive elements to engage your guests. This part also covers the importance of offering refreshments, fostering networking, and collecting feedback to ensure a memorable and successful art show.

11.3 Collaborating with Other Cloud Spotters

Collaboration can greatly enhance your cloud spotting experience. This section offers guidance on how to work with other cloud enthusiasts to create impactful projects and community art.

- Group Projects: Learn how to organize and execute group projects, from identifying participants and planning the project to assigning roles and showcasing the results. This part emphasizes the importance of regular meetings, open communication, and celebrating the collaboration.

- Community Art: Explore how to define and execute community art projects, such as murals or public art installations. This part provides tips on securing support and resources, promoting the project, and organizing workshops and planning sessions to engage community members.

- Online Collaborations: Discover the potential of online platforms for collaboration. This part covers finding online communities, proposing collaborative projects, using collaboration tools, maintaining communication, and sharing and promoting the final project.

By following the comprehensive guidance provided in Chapter 11, readers can transform their cloud spotting observations into beautiful and meaningful art. Whether working individually or with others, these projects and collaborations will deepen your appreciation for clouds and enhance your creative expression.

Part 6: Exploring Rare and Extraordinary Clouds

Chapter 12: Chasing Rare Cloud Formations

Cloud spotting can be incredibly rewarding when you encounter rare and extraordinary cloud formations. This chapter will guide you through identifying and photographing these unique clouds, share personal stories from extreme cloud spotting adventures, and explain the optical phenomena associated with clouds, such as halos, sundogs, and rainbows.

12.1 Identifying and Photographing Unique Clouds

Cloud spotting can be an exciting and rewarding hobby, especially when you encounter rare and extraordinary cloud formations. This section will guide you through identifying different types of rare clouds, provide essential photography tips to capture their beauty, and offer documentation techniques to record your findings effectively.

12.1.1 Rare Cloud Types

Mammatus Clouds

- Identification: Look for pouch-like structures hanging from the cloud base, typically associated with cumulonimbus clouds. They often appear after severe thunderstorms.

- Characteristics: Smooth, rounded lobes that can vary in size. They often look like a field of upside-down bubbles.

Lenticular Clouds

- Identification: These lens-shaped clouds often form near mountain ranges due to orographic lift. They are smooth, oval, and sometimes stacked in layers.

- Characteristics: Stationary despite strong winds, they are often mistaken for UFOs due to their saucer-like appearance.

Noctilucent Clouds

- Identification: Visible during twilight, these high-altitude clouds glow with a bluish-white light. They are found at altitudes around 76 to 85 kilometers in the mesosphere.

- Characteristics: Thin, wispy, and often appear as delicate, glowing filaments.

Kelvin-Helmholtz Clouds

- Identification: Resemble ocean waves with a rolling, wave-like appearance. Formed due to wind shear between different layers of air.

- Characteristics: Curved, cresting shapes that appear as if the clouds are breaking like waves on a beach.

Polar Stratospheric Clouds

- Identification: Found in polar regions during winter, these iridescent clouds are often called "mother-of-pearl" clouds.

- Characteristics: Bright, colorful displays due to sunlight reflecting off ice crystals.

12.1.2 Photography Tips

Equipment

- Camera: A DSLR or mirrorless camera with a high-quality zoom lens is ideal. A smartphone with a good camera can also be used, but ensure it has manual settings.

- Tripod: Essential for stability, especially in low light conditions or when using a slow shutter speed.

Camera Settings

- ISO: Use a low ISO (100-200) to minimize noise and maintain image clarity.

- Aperture: A small aperture (high f-stop, such as f/8 to f/16) will provide a greater depth of field, keeping more of the scene in focus.

- Shutter Speed: Adjust based on lighting conditions. For fast-moving clouds, a faster shutter speed (1/250s or faster) can freeze the motion. For a more dramatic effect with blurred motion, use a slower shutter speed (1/30s or slower).

Composition Techniques

- Rule of Thirds: Position the clouds along the grid lines or intersections to create a balanced and visually appealing composition.

- Foreground Elements: Include elements such as trees, mountains, or buildings to provide context and scale.

- Leading Lines: Use natural lines, such as roads or rivers, to draw the viewer's eye towards the clouds.

Lighting Conditions

- Golden Hours: The hours just after sunrise and just before sunset provide the best natural light, casting warm tones and creating long shadows.

- Twilight: For noctilucent clouds, twilight is the optimal time to capture their ethereal glow.

12.1.3 Documentation Techniques

Keeping a Log

- Details to Record: Date, time, location, weather conditions, and cloud type. Include any relevant notes on cloud movement, changes, and surrounding environment.

- Format: Use a dedicated cloud spotting journal or a digital app to organize your entries. Consistency is key for effective documentation.

Using Technology

- Apps: Utilize cloud identification apps such as CloudSpotter, MyRadar, or Weather Underground. These apps can help identify cloud types and track weather patterns.

- GPS Tagging: Use GPS to tag the location of your sightings. This can be especially useful for revisiting spots or sharing precise locations with other cloud enthusiasts.

Sharing Your Findings

- Online Communities: Join platforms like the Cloud Appreciation Society or local weather forums to share your observations and photographs. Engage with other cloud spotters to exchange tips and experiences.

- Social Media: Use hashtags related to cloud spotting (#cloudspotting, #clouds, #skywatching) to reach a broader audience. Share your best photos and stories to inspire others.

By mastering the identification of rare cloud types, using effective photography techniques, and meticulously documenting your observations, you can enhance your cloud spotting experience and contribute valuable information to the cloud spotting community.

12.2 Stories from the Field: Extreme Cloud Spotting Adventures

Cloud spotting often leads to incredible adventures and unforgettable experiences. This section delves into personal accounts of extreme cloud spotting, highlights memorable experiences, and shares valuable lessons learned from those who have ventured far and wide in pursuit of unique cloud formations.

12.2.1 Personal Accounts

Story 1: The Pursuit of Lenticular Clouds

- Background: A cloud spotter's journey to the Andes Mountains in South America.

- Experience: After weeks of planning and tracking weather patterns, the spotter finally captures a breathtaking view of stacked lenticular clouds over the mountain peaks.

- Challenges: Harsh weather conditions, high altitude, and unpredictable cloud appearances.

- Outcome: The spotter captures stunning photographs and gains a deep appreciation for the natural phenomena.

Story 2: Chasing Noctilucent Clouds in Scandinavia

- Background: A summer trip to the high latitudes of Scandinavia to observe noctilucent clouds.

- Experience: Long nights spent under the twilight sky, waiting for the elusive glow of these high-altitude clouds.

- Challenges: Cold temperatures, extended periods of daylight, and the need for precise timing.

- Outcome: Witnessing the ethereal beauty of noctilucent clouds and capturing their delicate glow on camera.

Story 3: The Hunt for Kelvin-Helmholtz Waves

- Background: A dedicated cloud spotter travels across the Great Plains of the United States.

- Experience: Spotting and photographing the rare wave-like Kelvin-Helmholtz clouds, often preceding severe weather.

- Challenges: Navigating through regions with volatile weather conditions and ensuring personal safety.

- Outcome: Capturing one of the most elusive and dynamic cloud formations, contributing valuable images to the cloud spotting community.

12.2.2 Memorable Experiences

Experience 1: Witnessing a Supercell Formation

- Location: Tornado Alley, USA.

- Event: A cloud spotter's experience during storm-chasing season, observing the development of a massive supercell thunderstorm.

- Details: The spotter describes the intense build-up of the storm, the dramatic cloud formations, and the raw power of nature.

- Takeaway: The awe-inspiring sight of a supercell and the importance of respecting nature's power.

Experience 2: A Journey to the Arctic for Polar Stratospheric Clouds

- Location: Svalbard, Norway.

- Event: An expedition to the Arctic Circle to witness the vibrant colors of polar stratospheric clouds.

- Details: The spotter endures extreme cold and long periods of darkness to capture the iridescent display of these rare clouds.

- Takeaway: The magical and surreal beauty of the Arctic sky and the reward of perseverance.

Experience 3: Capturing a Full-Circle Rainbow

- Location: Victoria Falls, Zambia/Zimbabwe border.

- Event: A cloud spotter's experience of capturing a rare full-circle rainbow from the vantage point of a high cliff.

- Details: The spotter describes the perfect combination of mist, sunlight, and elevation needed to observe this phenomenon.

- Takeaway: The joy of witnessing and capturing a full-circle rainbow and the importance of being at the right place at the right time.

12.2.3 Lessons Learned

Lesson 1: Preparation is Key

- Research: Thoroughly research your target cloud types, ideal locations, and best times for observation.

- Equipment: Ensure you have the right equipment, including cameras, lenses, weather-appropriate clothing, and safety gear.

- Weather Tracking: Use weather apps and forecasts to plan your outings and increase your chances of success.

Lesson 2: Patience and Persistence Pay Off

- Timing: Some of the most extraordinary cloud sightings require long hours of waiting and watching.

- Resilience: Be prepared for setbacks, such as unfavorable weather conditions or missed opportunities.

- Dedication: Maintain a positive attitude and stay committed to your passion for cloud spotting.

Lesson 3: Safety First

- Awareness: Always be aware of your surroundings, especially when dealing with extreme weather conditions.

- Safety Gear: Carry essential safety gear, such as first aid kits, communication devices, and emergency supplies.

- Risk Management: Know when to prioritize safety over the pursuit of cloud spotting. No photograph is worth risking your well-being.

Lesson 4: The Value of Community

- Networking: Connect with other cloud spotters and meteorologists to share knowledge, tips, and experiences.

- Collaboration: Participate in group expeditions and collaborative projects to enhance your cloud spotting adventures.

- Support: Lean on the cloud spotting community for support, whether it's for encouragement, safety advice, or sharing the joy of a great sighting.

By immersing yourself in the stories, experiences, and lessons of seasoned cloud spotters, you can gain valuable insights and inspiration for your own cloud spotting adventures. Embrace the challenges, savor the successes, and always prioritize safety and preparation in your pursuit of rare and extraordinary clouds.

12.3 Understanding Optical Phenomena: Halos, Sundogs, Rainbows

Optical phenomena such as halos, sundogs, and rainbows add a magical element to cloud spotting. Understanding how these phenomena form, how to identify them, and how to capture them effectively can greatly enhance your cloud spotting experience. This section provides a detailed guide to the formation processes, identification tips, and techniques for photographing these beautiful atmospheric displays.

12.3.1 Formation Processes

Halos

- Formation: Halos are created by the refraction, reflection, and dispersion of light through ice crystals in cirrus or cirrostratus clouds. The shape and orientation of these crystals determine the type of halo observed.

- Types of Halos: The most common halos are the 22° halo, which forms a circle around the sun or moon, and the 46° halo, which is larger and rarer. Other types include tangent arcs, sun pillars, and Parry arcs.

Sundogs (Parhelia)

- Formation: Sundogs occur due to the refraction of sunlight through hexagonal ice crystals in the atmosphere. These crystals act like prisms, bending the light at specific angles.

- Appearance: Sundogs are bright spots that appear on either side of the sun, usually at a distance of about 22 degrees. They often have a reddish hue on the side nearest the sun and a bluish-white hue on the opposite side.

Rainbows

- Formation: Rainbows are formed by the refraction, reflection, and dispersion of light through water droplets in the atmosphere. Each droplet acts like a miniature prism, breaking the light into its constituent colors.

- Types of Rainbows: The primary rainbow forms a circular arc with red on the outer edge and violet on the inner edge. A secondary rainbow, which is fainter and has reversed colors, can sometimes be seen outside the primary arc.

12.3.2 Identification Tips

Halos

- Where to Look: Halos are typically seen around the sun or moon. Look for a bright, circular ring at a distance of 22 or 46 degrees from the light source.

- When to Look: Halos are more likely to be seen when the sky is partially covered with thin, high-altitude cirrus or cirrostratus clouds.

- Other Indicators: Pay attention to other halo types such as sun pillars, which appear as vertical columns of light, or tangent arcs, which are bright spots at the top or bottom of a halo.

Sundogs

- Where to Look: Sundogs are usually found to the left and right of the sun, about 22 degrees away.

- When to Look: The best time to observe sundogs is when the sun is low on the horizon, either in the morning or late afternoon.

- Other Indicators: Sundogs are often accompanied by a 22° halo, so look for a bright ring around the sun as well.

Rainbows

- Where to Look: Rainbows appear opposite the sun, so if the sun is in the east, look towards the west, and vice versa.

- When to Look: Rainbows are most commonly seen during or after a rain shower when the sun is low in the sky, typically in the early morning or late afternoon.

- Other Indicators: Look for a faint secondary rainbow outside the primary arc. The area between the primary and secondary rainbows, known as Alexander's band, is usually darker.

12.3.3 Capturing Phenomena

Photographing Halos

- Equipment: Use a wide-angle lens to capture the entire halo. A DSLR or mirrorless camera with manual settings is ideal.

- Settings: Set your camera to a low ISO (100-200) to reduce noise, a small aperture (f/8 to f/16) for greater depth of field, and adjust the shutter speed to balance the exposure.

- Composition: Position the sun or moon in the center of your frame, and include foreground elements for scale and context. Use a tripod to keep your camera steady, especially in low light conditions.

Photographing Sundogs

- Equipment: Similar to halos, a wide-angle lens works best. A polarizing filter can help enhance the colors.

- Settings: Use a low ISO and a small aperture. Adjust the shutter speed to capture the bright spots without overexposing the sun.

- Composition: Frame the sundogs with the sun in the center, and try to capture both sundogs in a single shot. Include elements like trees or buildings for added interest.

Photographing Rainbows

- Equipment: A wide-angle lens is essential to capture the full arc of the rainbow. A polarizing filter can intensify the colors.

- Settings: Use a low ISO, a small aperture, and a shutter speed that balances the exposure of the rainbow and the surrounding landscape.

- Composition: Position the rainbow in the frame with a strong foreground element to provide depth. Capture reflections of the rainbow in water bodies for a more dynamic composition.

Post-Processing Tips

- Enhancing Colors: Use photo editing software to adjust the saturation and contrast to make the colors of the halos, sundogs, or rainbows more vivid.

- Reducing Glare: If shooting directly into the sun, reduce glare and enhance details using the dehaze tool in your editing software.

- Sharpening: Apply gentle sharpening to bring out the finer details in the optical phenomena without introducing noise.

By understanding the formation processes, identifying tips, and capturing techniques, you can successfully photograph and appreciate the beauty of halos, sundogs, and rainbows. These phenomena not only enhance your cloud spotting experience but also provide stunning visual rewards that can be shared and celebrated.

Chapter 13: Cloud Spotting Around the World

Cloud spotting is a universal hobby that offers unique experiences in different parts of the world. This chapter explores famous cloud spotting locations globally, provides essential travel tips for cloud enthusiasts, and explains how to connect with international cloud spotting communities.

13.1 Famous Cloud Spotting Locations Globally

Cloud spotting is a fascinating hobby that can take you to some of the most beautiful and unique places on Earth. This section highlights famous cloud spotting locations globally, providing details on what makes each location special and what types of clouds you can expect to see.

13.1.1 North America

Mount Rainier, USA

- Why It's Famous: Mount Rainier is renowned for its stunning lenticular clouds, which form due to the mountain's unique shape and weather patterns.

- What to Expect: Look for lens-shaped clouds that resemble flying saucers. These clouds are most commonly seen during the fall and winter months.

Tornado Alley, USA

- Why It's Famous: This region is known for its dramatic supercells and mammatus clouds, especially during the storm season in spring and early summer.

- What to Expect: Experience towering cumulonimbus clouds and the distinctive pouch-like formations of mammatus clouds. Storm chasing tours are available for those seeking an adrenaline-filled adventure.

The Great Plains, USA

- Why It's Famous: The vast, open skies of the Great Plains offer perfect conditions for observing a variety of cloud types, including expansive stratocumulus layers.

- What to Expect: Spot large, layered clouds that stretch across the horizon. The best times to visit are during the spring and summer months.

Rocky Mountains, Canada

- Why It's Famous: The Rocky Mountains provide excellent opportunities for observing orographic clouds, including lenticular and wave clouds.

- What to Expect: Look for clouds that form as air flows over the mountains, creating spectacular shapes and patterns. These clouds are often seen year-round but are particularly striking in the winter.

13.1.2 Europe

The Scottish Highlands, UK

- Why It's Famous: Known for its rapidly changing weather, the Highlands offer a dynamic cloud spotting experience with rare formations.

- What to Expect: Observe lenticular clouds and dramatic cumulonimbus formations. The best times to visit are during the fall and spring.

The Alps, Switzerland

- Why It's Famous: The high altitude and diverse weather patterns in the Alps make it a prime location for cloud spotting.

- What to Expect: Spot altostratus and cirrostratus clouds against a stunning mountainous backdrop. Winter and spring are the best seasons for cloud spotting here.

Norway's Coastline

- Why It's Famous: Norway's unique coastal geography provides excellent conditions for observing noctilucent clouds during the summer months.

- What to Expect: Look for glowing, high-altitude clouds visible during twilight. Visit between late May and early August for the best sightings.

The Baltic Sea, Sweden

- Why It's Famous: The Baltic Sea area is renowned for its beautiful stratocumulus and altocumulus clouds, especially during the long summer evenings.

- What to Expect: Experience layers of fluffy clouds illuminated by the setting sun. Summer months provide the best opportunities for cloud spotting.

13.1.3 Asia and Beyond

Mount Fuji, Japan

- Why It's Famous: Mount Fuji often attracts lenticular clouds, making it a favorite spot for cloud enthusiasts.

- What to Expect: Look for lens-shaped clouds forming over the mountain. The best time to visit is during the fall and winter months.

The Himalayas, Nepal

- Why It's Famous: The high altitudes and unique weather patterns of the Himalayas create opportunities to observe rare cloud types.

- What to Expect: Spot orographic and Kelvin-Helmholtz clouds. The best times to visit are during the spring and fall.

Outback, Australia

- Why It's Famous: The vast, open skies of the Australian Outback are perfect for spotting cirrus and cirrostratus clouds.

- What to Expect: Enjoy clear skies and minimal light pollution, making it ideal for observing noctilucent clouds. Visit during the winter months for the best experience.

Table Mountain, South Africa

- Why It's Famous: Known for its "tablecloth" cloud, a stunning orographic cloud that forms over the mountain.

- What to Expect: Observe a blanket-like cloud formation that spills over the mountain's edge. The best time to see this phenomenon is during the summer months.

By visiting these famous cloud spotting locations around the world, you can experience the beauty and diversity of cloud formations in different climates and landscapes. Whether you're drawn to the dramatic skies of Tornado Alley or the serene views over the Norwegian coast, these destinations offer unique and unforgettable cloud spotting experiences.

13.2 Travel Tips for Cloud Enthusiasts

Traveling for cloud spotting can be an enriching experience that combines your passion for clouds with the excitement of exploring new places. This section provides a comprehensive guide to planning your trips, packing essential gear, and considering local factors to ensure a successful and enjoyable cloud spotting adventure.

13.2.1 Planning Your Trip

Step 1: Research Destinations

- Identify Key Locations: Determine which locations are known for unique cloud formations and align with your interests.

- Best Times to Visit: Research the optimal times of year to visit each location. Consider weather patterns, seasonal changes, and local events that might impact your cloud spotting.

Step 2: Create an Itinerary

- Plan Your Route: Map out your route to include multiple cloud spotting sites. Ensure you have enough time at each location to observe and photograph clouds.

- Flexibility: Build flexibility into your schedule to accommodate changing weather conditions. Plan extra days or alternate activities in case of unfavorable weather.

Step 3: Accommodation and Transportation

- Booking: Reserve accommodations in advance, especially if traveling during peak seasons. Look for places close to your cloud spotting sites.

- Transport: Arrange for reliable transportation, whether renting a car, using public transport, or booking guided tours. Ensure your transportation can handle the local terrain and weather conditions.

Step 4: Permits and Regulations

- Check Requirements: Verify if any permits are required for accessing certain areas or national parks. Adhere to local regulations and guidelines.

- Safety Measures: Familiarize yourself with local safety protocols, especially if visiting remote or high-altitude locations.

13.2.2 Essential Gear

Step 1: Camera Equipment

- Camera: A DSLR or mirrorless camera with manual settings is ideal. A smartphone with a good camera can also be used.

- Lenses: Bring a variety of lenses, including wide-angle for landscapes and zoom for capturing details.

- Tripod: Essential for stable shots, especially in low light conditions or for long exposures.

Step 2: Weather Gear

- Clothing: Pack weather-appropriate clothing, including waterproof jackets, layered clothing, hats, and gloves. Dress in layers to adapt to changing temperatures.

- Footwear: Wear sturdy, comfortable shoes suitable for walking and hiking. Waterproof boots are recommended for wet or muddy conditions.

- Sun Protection: Bring sunglasses, sunscreen, and a hat to protect against UV rays.

Step 3: Safety and Navigation

- First Aid Kit: Include basic medical supplies such as bandages, antiseptics, and pain relievers.

- Navigation Tools: Carry maps, a compass, or a GPS device. A smartphone with offline maps can be very useful.

- Communication: Ensure you have a way to communicate in case of emergencies, such as a fully charged mobile phone or a satellite phone for remote areas.

Step 4: Additional Supplies

- Batteries and Memory Cards: Bring extra batteries and memory cards for your camera to avoid running out of power or storage.

- Backpack: Use a comfortable, weather-resistant backpack to carry your gear and supplies.

- Snacks and Water: Pack enough food and water to stay hydrated and energized during your outings.

13.2.3 Local Considerations

Step 1: Cultural Awareness

- Respect Local Customs: Be mindful of local traditions and cultural norms. Dress appropriately and follow local etiquette.

- Language: Learn basic phrases in the local language to help with navigation and communication. Use translation apps if necessary.

Step 2: Environmental Respect

- Leave No Trace: Follow the Leave No Trace principles to minimize your impact on the environment. Pack out all trash and avoid disturbing wildlife.

- Local Wildlife: Be aware of local wildlife and how to interact with or avoid them safely. Research any potentially dangerous animals in the area.

Step 3: Engaging with Locals

- Hire Local Guides: Consider hiring a local guide who knows the area well and can provide valuable insights and safe navigation to the best cloud spotting locations.

- Local Communities: Engage with local communities to learn more about the region and its natural beauty. Participate in local events or activities if possible.

Step 4: Safety Precautions

- Weather Conditions: Stay informed about current and forecasted weather conditions. Adjust your plans accordingly to avoid hazardous weather.

- Emergency Plans: Have a plan in place for emergencies, including the nearest medical facilities and emergency contact numbers.

By following these detailed travel tips, you can enhance your cloud spotting adventures, ensuring they are safe, enjoyable, and successful. Proper planning, packing the right gear, and being mindful of local considerations will help you make the most of your cloud spotting trips around the world.

13.3 Connecting with International Cloud Spotting Communities

Connecting with fellow cloud spotters around the world can enrich your cloud spotting experience by providing a platform to share knowledge, exchange ideas, and collaborate on projects. This section outlines how to engage with global networks, participate in online communities, and join collaborative events to enhance your cloud spotting journey.

13.3.1 Global Networks

Step 1: Join International Organizations

- Cloud Appreciation Society: Become a member of the Cloud Appreciation Society, a global community dedicated to the love of clouds. Membership provides access to a wealth of resources, including educational materials, forums, and events.

- Meteorological Societies: Join meteorological societies that often have dedicated sections or interest groups for cloud spotting. These societies can provide scientific insights and networking opportunities.

Step 2: Participate in Conferences and Workshops

- Attend Events: Look for international conferences, workshops, and seminars focused on meteorology and cloud spotting. These events are excellent for networking, learning, and sharing your own experiences.

- Present Your Findings: If you have significant observations or research, consider presenting at these events. Sharing your work can spark discussions and collaborations.

Step 3: Network with Experts

- Reach Out to Professionals: Connect with meteorologists, researchers, and experienced cloud spotters. Engaging with experts can provide deeper insights and advanced knowledge.

- Join Research Projects: Participate in citizen science projects or research initiatives that focus on cloud observation and weather phenomena.

13.3.2 Online Communities

Step 1: Find and Join Online Platforms

- Social Media Groups: Join cloud spotting groups on platforms like Facebook, Instagram, and Reddit. These groups often share stunning cloud photos, identification tips, and weather updates.

- Specialized Forums: Participate in forums such as Cloud Appreciation Society's forum, Weather Underground, and Met Office forums. These platforms are great for detailed discussions and technical advice.

Step 2: Engage with the Community

- Share Your Observations: Post your cloud photos, observations, and experiences. Engage with others by commenting, liking, and sharing their posts.

- Ask Questions: Don't hesitate to ask for help with cloud identification or advice on cloud spotting techniques. The community is often very supportive and knowledgeable.

Step 3: Participate in Online Events

- Webinars and Livestreams: Attend online webinars and livestreams hosted by cloud spotting groups and meteorological societies. These events often feature expert talks, Q&A sessions, and live cloud observations.

- Virtual Meetups: Join virtual meetups organized by cloud spotting communities. These meetups are a great way to connect with other enthusiasts from around the world.

Step 4: Utilize Technology for Collaboration

- Cloud Spotting Apps: Use cloud spotting apps like CloudSpotter, MyRadar, and Weather Underground to share your sightings and collaborate with others.

- Online Tools: Utilize tools like Google Drive, Dropbox, and collaborative platforms like Slack or Discord to work on joint projects and share data.

13.3.3 Collaborative Events

Step 1: Organize or Join Cloud Spotting Expeditions

- Group Expeditions: Organize or join group expeditions to famous cloud spotting locations. Collaborating on trips can enhance the experience and increase the chances of spotting unique cloud formations.

- Guided Tours: Participate in guided cloud spotting tours offered by travel companies or cloud spotting groups. These tours are led by experts and often provide access to prime cloud spotting sites.

Step 2: Participate in Collaborative Projects

- Citizen Science Initiatives: Join citizen science projects that involve cloud observation and data collection. Your contributions can help with scientific research and weather forecasting.

- Photo Contests and Exhibitions: Enter cloud photography contests or contribute to cloud-themed art exhibitions. These events are great for showcasing your work and seeing the creativity of others.

Step 3: Attend Global Cloud Spotting Festivals

- Cloud Festivals: Some regions host festivals dedicated to cloud spotting and weather phenomena. These festivals feature workshops, talks, and group spotting activities.

- Networking Opportunities: Use these festivals as an opportunity to network with other cloud enthusiasts, share your experiences, and learn from others.

Step 4: Engage in Educational Outreach

- Workshops and Lectures: Organize or participate in workshops and lectures at schools, community centers, or online. Educating others about cloud spotting can foster a broader appreciation for the hobby.

- Collaborate with Educational Institutions: Partner with schools and universities to develop cloud spotting programs or contribute to their meteorology courses.

By actively engaging with global networks, participating in online communities, and joining collaborative events, you can greatly enhance your cloud spotting experience. Connecting with other enthusiasts around the world not only broadens your knowledge but also provides a supportive community where you can share your passion for clouds.

Chapter 13 Review

Chapter 13 of the HowExpert Guide to Cloud Spotting offers a comprehensive guide to exploring cloud spotting on a global scale. This chapter provides valuable insights into famous cloud spotting locations worldwide, essential travel tips for cloud enthusiasts, and practical advice on connecting with international cloud spotting communities.

13.1 Famous Cloud Spotting Locations Globally

This section highlights some of the best places around the world for cloud spotting. It details specific locations in North America, Europe, and Asia and beyond, known for their unique and spectacular cloud formations. From the lenticular clouds of Mount Rainier to the noctilucent clouds over Norway's coastline, readers will discover the top destinations to experience extraordinary cloud formations. Each location is described with information on what makes it special and the types of clouds you can expect to see.

13.2 Travel Tips for Cloud Enthusiasts

Planning a cloud spotting trip requires careful preparation to ensure a successful and enjoyable experience. This section provides step-by-step guidance on planning your trip, including researching destinations, creating an itinerary, booking accommodations, and arranging transportation. It also covers essential gear to pack, such as camera equipment, weather-appropriate clothing, and safety supplies. Additionally, readers will find tips on respecting local cultures and environments, engaging with locals, and staying safe during their travels.

13.3 Connecting with International Cloud Spotting Communities

Engaging with other cloud enthusiasts worldwide can greatly enhance your cloud spotting journey. This section outlines how to connect with global networks, participate in online communities, and join collaborative events. Readers will learn about joining international organizations, attending conferences and workshops, and networking with experts. The section also provides information on utilizing social media groups, forums, and cloud spotting apps to share observations and participate in virtual events. Finally, it highlights the benefits of collaborative projects, citizen science initiatives, and educational outreach, encouraging readers to join group expeditions and participate in cloud-themed festivals.

By exploring the famous cloud spotting locations globally, following practical travel tips, and connecting with international communities, readers can elevate their cloud spotting experience. This chapter equips cloud enthusiasts with the knowledge and tools to explore new destinations, capture stunning cloud formations, and engage with a supportive community of fellow cloud spotters worldwide.

Part 7: The Future of Cloud Spotting

Chapter 14: Clouds and Climate Change

Understanding the relationship between clouds and climate change is essential for both cloud spotters and scientists. This chapter explores how changing climates affect cloud formation, the role of citizen science in cloud research, and innovations in cloud observation that are shaping the future of this field.

14.1 How Changing Climates Affect Clouds

Understanding how climate change impacts cloud formation is crucial for cloud spotters and scientists alike. This section delves into the effects of changing climates on cloud formation, regional variations, and long-term changes. By exploring these areas, you can gain a deeper appreciation of the intricate relationship between clouds and our planet's climate.

14.1.1 Impact on Cloud Formation

Step 1: Temperature Changes

- Warmer Temperatures: Increased global temperatures lead to higher evaporation rates, adding more moisture to the atmosphere. This can result in more cloud formation, particularly in tropical regions.

- Thermal Instability: Warmer temperatures can cause thermal instability in the atmosphere, promoting the development of cumulonimbus clouds, which are responsible for thunderstorms and heavy rainfall.

Step 2: Moisture Levels

- Increased Humidity: Higher levels of atmospheric moisture can lead to the formation of thicker and more widespread cloud cover. This is especially noticeable in regions experiencing increased precipitation.

- Drier Conditions: In contrast, areas experiencing drought or reduced humidity may see a decrease in cloud cover, leading to more clear skies and increased surface heating.

Step 3: Air Circulation Patterns

- Jet Stream Shifts: Changes in global air circulation patterns, such as shifts in the jet stream, can alter where and how clouds form. These shifts can result in more frequent and intense weather events in certain regions.

- Wind Patterns: Alterations in wind patterns can affect the distribution and types of clouds observed. For example, strong winds can spread moisture over larger areas, promoting cloud formation.

14.1.2 Regional Variations

Step 1: Polar Regions

- Arctic and Antarctic: Warming temperatures in polar regions are causing sea ice to melt, increasing atmospheric moisture. This can lead to more low-level cloud cover, which can trap heat and contribute to further warming.

- Cloud Types: Expect to see more stratiform clouds, which are extensive and layered, reflecting some sunlight but also trapping heat.

Step 2: Tropical Regions

- Sea Surface Temperatures: Rising sea surface temperatures enhance convection processes, leading to the formation of more cumulus and cumulonimbus clouds.

- Weather Patterns: These regions may experience more intense tropical storms and hurricanes, characterized by towering cumulonimbus clouds and extensive cirrus clouds at higher altitudes.

Step 3: Mid-latitude Regions

- Storm Frequency: Changes in precipitation patterns and the frequency of storm systems can alter cloud cover. Increased storm activity can lead to more persistent cloud formations, such as altostratus and nimbostratus clouds.

- Seasonal Variations: Seasonal changes in temperature and humidity will also affect cloud types and coverage, with more dynamic cloud formations during transitional seasons like spring and autumn.

14.1.3 Long-term Changes

Step 1: Cloud Cover Trends

- Satellite Data: Long-term satellite observations reveal trends in cloud cover changes over decades. Some regions may see an increase in cloudiness due to higher evaporation rates, while others may experience a decrease due to changing weather patterns.

- Climate Models: Climate models predict varying impacts on cloud cover based on different greenhouse gas emission scenarios. Understanding these models helps in forecasting future cloud patterns.

Step 2: Cloud Reflectivity

- Albedo Effect: Changes in cloud composition and altitude can affect their reflectivity (albedo). Higher and thinner clouds, such as cirrus clouds, may allow more sunlight to reach the Earth's surface, contributing to warming.

- Radiative Forcing: The balance between cloud cooling and warming effects is crucial in understanding their impact on global temperatures. Increased low-level cloud cover can enhance cooling, while high-altitude clouds can contribute to warming.

Step 3: Feedback Mechanisms

- Positive Feedback: Increased cloud cover can trap more heat in the atmosphere, leading to a positive feedback loop that accelerates global warming. This is particularly significant with clouds that have high water vapor content.

- Negative Feedback: Conversely, clouds that reflect more sunlight back into space can have a cooling effect, creating a negative feedback loop that helps to mitigate warming. Stratus clouds over oceans are an example of this effect.

By comprehensively understanding how changing climates affect cloud formation, regional variations, and long-term changes, you can better appreciate the dynamic nature of clouds and their crucial role in the Earth's climate system. This knowledge is essential for cloud spotters, as it enhances the ability to predict and observe cloud patterns in a changing world.

14.2 The Role of Citizen Science in Cloud Research

Citizen science plays a vital role in cloud research by enabling enthusiasts and volunteers to contribute valuable data and observations. This section outlines the opportunities for contribution, methods of data collection, and the ways in which citizen science influences research. By participating in these activities, cloud spotters can help advance our understanding of clouds and their impact on the climate.

14.2.1 Contribution Opportunities

Step 1: Join Cloud Observation Programs

- NASA's GLOBE Program: Participate in the GLOBE Program, which allows volunteers to record and share cloud observations. This data helps scientists understand cloud patterns and their effects on the climate.

- Citizen Science Platforms: Explore platforms like Zooniverse, where various projects need cloud data. Join relevant projects to contribute your observations.

Step 2: Share Cloud Photographs

- Online Databases: Contribute your cloud photographs to online databases such as the Cloud Appreciation Society or weather-focused platforms like Weather Underground. These images can be used by researchers to study cloud formations and trends.

- Social Media: Use hashtags like #CloudSpotting, #Clouds, and #CitizenScience to share your photos on social media platforms. Tagging organizations and using specific project hashtags can increase the visibility and use of your contributions.

Step 3: Participate in Events and Workshops

- Local Events: Attend local cloud spotting events, workshops, and training sessions. These gatherings provide hands-on opportunities to learn about cloud observation and contribute to larger data collection efforts.

- Online Webinars: Join online webinars and virtual events hosted by meteorological societies and citizen science organizations to stay updated on new projects and methods.

14.2.2 Data Collection

Step 1: Utilize Cloud Observation Apps

- CloudSpotter App: Download and use the CloudSpotter app, which helps you identify cloud types and submit your observations to a global database. The app guides you through the process and ensures data accuracy.

- GLOBE Observer App: Use NASA's GLOBE Observer app to submit your cloud observations. The app includes tutorials and tools to help you collect high-quality data.

Step 2: Maintain Consistent Logs

- Observation Logbook: Keep a detailed logbook of your cloud observations, including date, time, location, cloud types, and weather conditions. Consistency in logging provides valuable long-term data for researchers.

- Digital Logs: Consider maintaining digital logs using spreadsheets or specialized software that can easily be shared and analyzed by researchers.

Step 3: Use Basic Meteorological Instruments

- Thermometers and Hygrometers: Complement your cloud observations with temperature and humidity data using basic instruments. This additional data provides context for your cloud observations.

- Barometers: Measure atmospheric pressure to understand the weather conditions associated with different cloud types.

Step 4: Take Detailed Photographs

- Photography Techniques: Use proper photography techniques to capture clear and detailed images of clouds. Ensure your camera settings are optimized for different lighting conditions and cloud types.

- Metadata: Include metadata such as date, time, and location in your photographs. This information is crucial for researchers analyzing the images.

14.2.3 Influencing Research

Step 1: Analyze and Interpret Data

- Basic Analysis: Learn how to analyze your cloud data using basic statistical tools and visualization techniques. This helps you understand trends and patterns in your observations.

- Citizen Science Platforms: Use tools provided by citizen science platforms to analyze and visualize your data. Many platforms offer user-friendly interfaces for data analysis.

Step 2: Share Your Findings

- Online Communities: Share your analyzed data and findings with online communities and forums. Engaging with other cloud spotters and researchers can lead to new insights and collaborations.

- Research Publications: Consider writing articles or reports based on your data and submitting them to citizen science journals or newsletters. Sharing your findings can contribute to the broader scientific community.

Step 3: Advocate and Educate

- Public Presentations: Give presentations at local schools, community centers, or cloud spotting events to share your knowledge and findings. Educating others about cloud research and its importance can inspire more people to participate.

- Social Media Campaigns: Use social media to advocate for citizen science and climate research. Share interesting findings, facts about clouds, and the importance of public contributions to science.

Step 4: Engage with Professional Researchers

- Collaborate on Projects: Reach out to professional researchers and offer to collaborate on projects. Your data and observations can complement their research and lead to significant discoveries.

- Contribute to Papers: Co-author research papers with scientists who use your data. This can provide recognition for your contributions and further integrate citizen science into professional research.

By participating in citizen science initiatives, maintaining detailed logs, and sharing your findings, you can significantly contribute to cloud research. Your efforts help scientists understand cloud behaviors, their impact on the climate, and how to better predict weather patterns. Engage with the community, utilize technology, and embrace the role of citizen science in advancing our knowledge of the skies.

14.3 Innovations in Cloud Observation

Cloud observation is a continually evolving field, benefiting from advancements in technology and new trends in research and collaboration. This section explores the latest technologies, emerging trends, and future prospects in cloud observation. By staying informed about these innovations, cloud spotters can enhance their skills and contribute more effectively to the field.

14.3.1 New Technologies

Step 1: Satellite Imaging

- High-Resolution Imaging: Modern satellites provide high-resolution images that capture intricate details of cloud formations. Instruments like the MODIS (Moderate Resolution Imaging Spectroradiometer) on NASA's Terra and Aqua satellites offer daily global coverage.

- Real-Time Data: Satellites like GOES (Geostationary Operational Environmental Satellites) provide near real-time data, allowing for continuous monitoring of cloud dynamics and weather systems.

Step 2: Drones and UAVs

- Data Collection: Unmanned Aerial Vehicles (UAVs) equipped with advanced sensors can collect data on cloud structure, temperature, and moisture content at various altitudes. This allows for detailed, localized studies that are not possible with satellites alone.

- Remote Areas: Drones can access remote or hazardous areas that are difficult or dangerous for humans to reach, providing valuable data on cloud formations in these regions.

Step 3: Ground-Based Lidar and Radar

- Lidar Technology: Light Detection and Ranging (Lidar) systems use lasers to measure cloud height, density, and composition. Ground-based Lidar can provide high-resolution vertical profiles of clouds.

- Doppler Radar: Doppler radar systems track cloud movement and precipitation, offering insights into storm development and progression. This technology is essential for understanding cloud dynamics and weather forecasting.

Step 4: Machine Learning and AI

- Data Analysis: Machine learning algorithms analyze vast amounts of cloud data, identifying patterns and making predictions about cloud behavior. AI can process data from multiple sources, providing more accurate and comprehensive insights.

- Automated Classification: AI systems can automatically classify cloud types from images, increasing the efficiency and accuracy of cloud observations. This is particularly useful for large datasets collected from satellites and ground-based observations.

14.3.2 Emerging Trends

Step 1: High-Altitude Observations

- Cirrus Clouds: Research is increasingly focusing on high-altitude cirrus clouds, which play a significant role in the Earth's radiation budget. Understanding their formation and behavior is crucial for climate models.

- Stratospheric Clouds: Studies on polar stratospheric clouds, which impact ozone depletion, are becoming more prevalent with advances in high-altitude research technology.

Step 2: Integrated Observation Systems

- Multi-Source Data Integration: Combining data from satellites, ground-based observations, and UAVs provides a more comprehensive view of cloud dynamics. Integrated systems offer real-time data and improve the accuracy of weather and climate models.

- Global Observation Networks: Collaborative networks like the Global Atmospheric Watch (GAW) integrate observations from multiple countries, enhancing global cloud monitoring and research.

Step 3: Citizen Science Integration

- Crowdsourced Data: Platforms that incorporate citizen science data with professional research are growing. This integration ensures that citizen-collected data is validated and useful for scientific studies.

- Public Engagement: Increasing public participation in cloud observation through educational programs and mobile apps encourages widespread data collection and enhances the scope of cloud research.

14.3.3 Future Prospects

Step 1: Enhanced Predictive Models

- Improved Forecasting: With better data from advanced technologies and integrated observation systems, predictive models of cloud behavior and climate interactions are becoming more accurate. This leads to improved weather forecasting and climate predictions.

- Climate Models: Enhanced models incorporating detailed cloud dynamics can better predict long-term climate changes and their impacts on various regions.

Step 2: Global Collaboration

- International Projects: Increased global collaboration among scientists, meteorologists, and citizen scientists is fostering a more comprehensive approach to cloud and climate research. Collaborative projects and data sharing platforms are enhancing collective analysis and understanding.

- Standardized Protocols: Developing standardized protocols for data collection and analysis ensures consistency and reliability across different regions and research groups.

Step 3: Public Engagement and Education

- Educational Outreach: Engaging the public in cloud observation and climate science is essential for raising awareness and driving action on climate issues. Future initiatives will focus on making cloud spotting more accessible and impactful for people of all ages.

- Interactive Platforms: Development of interactive platforms and virtual reality experiences can provide immersive learning opportunities about cloud dynamics and weather systems.

By embracing new technologies, staying informed about emerging trends, and participating in global collaboration, cloud spotters can significantly contribute to the advancement of cloud observation. These innovations not only enhance individual cloud spotting experiences but also play a crucial role in our understanding of the Earth's climate system and its future.

Chapter 14 Review

Chapter 14 of the HowExpert Guide to Cloud Spotting explores the profound relationship between clouds and climate change. This chapter provides readers with an in-depth understanding of how climate changes affect cloud formation, the vital role of citizen science in cloud research, and the latest innovations in cloud observation. By delving into these topics, readers will gain valuable insights into the dynamic nature of clouds and their critical role in the Earth's climate system.

14.1 How Changing Climates Affect Clouds

This section examines the various ways climate change impacts cloud formation. Readers will learn about:

- Impact on Cloud Formation: How rising temperatures, altered moisture levels, and shifting air circulation patterns influence the development and types of clouds.

- Regional Variations: The specific effects of climate change on cloud formations in polar, tropical, and mid-latitude regions, highlighting unique changes in each area.

- Long-term Changes: Trends in cloud cover and reflectivity, the role of clouds in climate feedback mechanisms, and how these changes are tracked over decades through satellite data and climate models.

14.2 The Role of Citizen Science in Cloud Research

Citizen science is crucial in enhancing our understanding of clouds and their role in climate science. This section covers:

- Contribution Opportunities: Ways for enthusiasts to join cloud observation programs, share photographs, and participate in workshops and events.

- Data Collection: Techniques for maintaining consistent observation logs, using cloud observation apps, and employing basic meteorological instruments to gather comprehensive data.

- Influencing Research: The impact of citizen-collected data on scientific research, methods for analyzing and sharing findings, and opportunities to collaborate with professional researchers.

14.3 Innovations in Cloud Observation

Technological advancements are revolutionizing cloud observation. This section explores:

- New Technologies: High-resolution satellite imaging, the use of drones and UAVs for data collection, ground-based Lidar and radar, and the application of machine learning for data analysis.

- Emerging Trends: Focus on high-altitude cloud studies, integrated observation systems combining multiple data sources, and the increasing role of citizen science in professional research.

- Future Prospects: The potential for enhanced predictive models, the importance of global collaboration in cloud research, and initiatives to engage the public in cloud observation and climate science education.

By exploring the comprehensive insights provided in Chapter 14, readers can enhance their understanding of the impact of climate change on clouds, actively contribute to cloud research through citizen science, and stay updated with the latest technological innovations in cloud observation. This chapter equips cloud enthusiasts with the knowledge and tools to play a vital role in advancing cloud and climate science.

Chapter 15: Inspiring Future Cloud Spotters

Cloud spotting is a fascinating and educational hobby that can inspire people of all ages. This chapter provides a comprehensive guide to teaching cloud spotting to kids, creating cloud spotting clubs and events, and building a community of cloud enthusiasts. By following these steps, you can foster a love for cloud spotting in future generations and create a supportive network of cloud spotters.

15.1 Teaching Cloud Spotting to Kids

Introducing children to cloud spotting can ignite their curiosity about the natural world and foster a lifelong interest in science and meteorology. This section provides a comprehensive guide to engaging kids in cloud spotting through educational activities, fun projects, and strategies for building sustained interest.

15.1.1 Educational Activities

Step 1: Cloud Identification Games

1. Flashcards: Create or purchase cloud identification flashcards featuring different cloud types and their characteristics.

2. Interactive Apps: Use cloud identification apps that include games and quizzes to help kids learn about various cloud formations.

3. Cloud Posters: Hang cloud posters in classrooms or kids' rooms as a visual aid to help them recognize and name different clouds.

Step 2: Cloud Watching Sessions

1. Observation Tools: Provide binoculars and cloud identification guides to enhance the observation experience.

2. Discussion Circles: After observing, gather the kids to discuss the types of clouds they saw, the weather conditions, and what they learned.

3. Weather Journals: Encourage kids to keep a weather journal where they can draw pictures of clouds they spot and write descriptions of their observations.

Step 3: Weather Journals

1. Daily Entries: Have kids record daily entries in their weather journals, noting the types of clouds they see, the weather conditions, and any changes over time.

2. Creative Additions: Encourage them to add creative elements like drawings, poems, or stories about their cloud spotting experiences.

3. Reflection: Periodically review the journal entries to discuss patterns and changes in the weather, reinforcing observational skills and curiosity.

15.1.2 Fun Projects

Step 1: Cloud Art

1. Materials: Provide various art supplies such as cotton balls, tissue paper, paint, and construction paper.

2. Projects: Organize projects where kids create their own cloud art. For example, use cotton balls to represent cumulus clouds or tissue paper for wispy cirrus clouds.

3. Exhibit: Display the completed artwork in a "Cloud Gallery" at home or school to celebrate their creativity.

Step 2: Cloud in a Jar

1. Materials: Gather a jar, hot water, ice, and aerosol spray (like hairspray).

2. Procedure: Pour hot water into the jar, place a plate with ice on top, and spray aerosol into the jar to form a cloud. Explain the science behind the cloud formation process.

3. Discussion: Discuss how clouds form in the atmosphere, relating the experiment to real-life cloud formation.

Step 3: Cloud Photography

1. Basics: Teach kids the basics of photography, including how to hold a camera, frame a shot, and use different settings.

2. Photo Walks: Organize photo walks where kids can take pictures of clouds. Encourage them to capture different types of clouds and weather conditions.

3. Photo Album: Create a cloud photo album or digital gallery where kids can display and share their best cloud photographs.

15.1.3 Building Interest

Step 1: Storytelling

1. Cloud Myths and Legends: Share stories and myths about clouds from different cultures. For example, tell the story of how the Greeks believed clouds were the work of Zeus.

2. Books and Movies: Read books or watch movies that feature clouds or weather as central themes. Discuss the scientific aspects and how they relate to what the kids have learned.

3. Personal Stories: Share your own cloud spotting experiences and why you enjoy it. Personal enthusiasm can be very contagious.

Step 2: Field Trips

1. Planning: Organize field trips to locations with interesting cloud formations, such as mountains, lakes, or parks. Research the best times and weather conditions for optimal cloud spotting.

2. Observation: During the trip, guide kids in observing and identifying different cloud types. Provide binoculars and cloud charts for hands-on learning.

3. Reflection: After the trip, have a discussion or a creative session where kids can share their experiences and what they learned.

Step 3: Guest Speakers

1. Inviting Experts: Invite meteorologists, climatologists, or experienced cloud spotters to talk to the kids about their work and passion for clouds.

2. Interactive Sessions: Encourage interactive sessions where kids can ask questions and engage with the speaker.

3. Inspiration: Highlight how the speaker's work relates to cloud spotting and the broader field of meteorology, inspiring kids to explore these areas further.

By implementing these educational activities, fun projects, and interest-building strategies, you can successfully engage kids in cloud spotting. This not only nurtures their curiosity and observational skills but also lays the foundation for a lifelong appreciation of the natural world and scientific inquiry.

15.2 Creating Cloud Spotting Clubs and Events

Creating a cloud spotting club and organizing events can foster a community of like-minded enthusiasts and provide a platform for sharing knowledge and experiences. This section outlines step-by-step instructions for starting a cloud spotting club, organizing events, and engaging members to ensure long-term success and participation.

15.2.1 Starting a Club

Step 1: Define the Club's Purpose

1. Mission Statement: Develop a clear mission statement that outlines the goals and objectives of the club, such as promoting cloud spotting, educating members about meteorology, and organizing regular observation sessions.

2. Target Audience: Identify your target audience. Will your club be open to all ages, or will it focus on specific groups like students, families, or retirees?

Step 2: Recruit Members

1. Spread the Word: Use social media, community bulletin boards, schools, and local organizations to spread the word about your new club. Create flyers and posters to attract potential members.

2. Introductory Meeting: Host an introductory meeting to provide information about the club, its goals, and planned activities. Use this opportunity to gather feedback and ideas from interested individuals.

Step 3: Establish a Structure

1. Leadership Team: Form a leadership team to help manage the club. Assign roles such as president, vice president, secretary, and treasurer to ensure smooth operation.

2. Meeting Schedule: Decide on a regular meeting schedule (e.g., monthly or bi-weekly) and choose convenient times and locations for members to gather.

Step 4: Create Guidelines and Policies

1. Membership Rules: Develop guidelines for membership, including any dues or fees, expected behavior, and participation requirements.

2. Safety Policies: Establish safety policies for outdoor activities and events, including weather considerations and emergency procedures.

15.2.2 Organizing Events

Step 1: Plan Regular Cloud Spotting Walks

1. Choose Locations: Select locations with good visibility and interesting cloud formations, such as parks, hilltops, or open fields.

2. Preparation: Prepare cloud identification guides, binoculars, and cameras for the walks. Provide members with a brief overview of what to look for and how to record their observations.

3. Guided Sessions: During the walks, offer guidance on identifying different cloud types and photographing them. Encourage members to share their observations and ask questions.

Step 2: Host Workshops and Lectures

1. Guest Speakers: Invite meteorologists, climatologists, and experienced cloud spotters to give talks and workshops. This provides members with expert insights and knowledge.

2. Hands-On Activities: Organize hands-on activities, such as cloud photography tutorials, cloud chart creation, and DIY weather station building.

Step 3: Organize Photo Contests and Exhibitions

1. Photo Contests: Hold cloud photography contests to encourage members to capture and share their best cloud photos. Offer small prizes or certificates for the best entries.

2. Exhibitions: Set up exhibitions of members' cloud photographs at local libraries, community centers, or online galleries. This showcases the club's activities and attracts new members.

Step 4: Plan Special Events

1. Field Trips: Organize field trips to famous cloud spotting locations, such as mountains, lakes, or coastal areas. These trips can provide unique cloud spotting opportunities and foster camaraderie among members.

2. Seasonal Celebrations: Plan seasonal events, like a summer solstice cloud spotting picnic or a winter sky observation night, to keep the club active year-round.

15.2.3 Engaging Members

Step 1: Interactive Activities

1. Cloud Quizzes: Create cloud quizzes and trivia games to test members' knowledge in a fun and engaging way.

2. Scavenger Hunts: Organize cloud scavenger hunts where members search for specific cloud types or weather phenomena.

Step 2: Communication Channels

1. Email Newsletters: Send regular email newsletters to keep members informed about upcoming events, club news, and interesting cloud-related articles or resources.

2. Social Media Groups: Create private social media groups for members to share their cloud photos, discuss observations, and stay connected between meetings.

Step 3: Recognition and Awards

1. Certificates and Badges: Recognize active members with certificates of participation, achievement badges, or small awards. This can motivate continued involvement and contributions.

2. Member Spotlights: Feature a different member each month in the newsletter or on social media. Highlight their contributions, cloud photos, and personal cloud spotting experiences.

Step 4: Continuous Learning Opportunities

1. Resource Library: Build a resource library of books, articles, and online tutorials on cloud spotting and meteorology. Make these resources available to all members.

2. Skill-Building Workshops: Offer workshops on advanced topics like cloud dynamics, weather forecasting, and climate science to deepen members' understanding and skills.

By following these detailed steps, you can successfully create a cloud spotting club, organize engaging events, and keep members actively involved. These efforts will help build a vibrant community of cloud enthusiasts who share a passion for observing and understanding the skies.

15.3 Building a Community of Cloud Enthusiasts

Creating and nurturing a community of cloud enthusiasts can significantly enhance the cloud spotting experience. This section provides a comprehensive guide to building a vibrant community through effective networking, leveraging online platforms, and ensuring long-term engagement.

15.3.1 Networking Tips

Step 1: Organize Local Meetups

1. Regular Gatherings: Schedule regular meetups at local parks, cafes, or community centers where cloud enthusiasts can share their experiences and learn from each other.

2. Theme-Based Events: Plan meetups around specific themes, such as cloud photography, weather forecasting, or seasonal cloud changes, to attract diverse participants.

3. Inviting Experts: Invite local meteorologists or experienced cloud spotters to speak at your meetups, providing valuable insights and fostering connections.

Step 2: Collaborate with Other Groups

1. Partnerships: Form partnerships with local schools, environmental organizations, and meteorological societies to broaden your reach and engage more people.

2. Joint Events: Co-host events with other groups, such as nature walks, science fairs, or environmental workshops, to introduce cloud spotting to new audiences.

3. Resource Sharing: Share resources and expertise with partner organizations to enhance the quality of your events and activities.

Step 3: Attend Conferences and Festivals

1. Relevant Conferences: Attend conferences and festivals focused on weather, meteorology, and environmental science to network with professionals and enthusiasts.

2. Networking Strategies: Prepare business cards or flyers about your cloud spotting community to distribute at these events. Engage in conversations and exchange contact information with attendees.

3. Follow-Up: After attending events, follow up with new contacts via email or social media to build and maintain connections.

15.3.2 Online Platforms

Step 1: Social Media Groups

1. Create a Group: Establish a dedicated social media group on platforms like Facebook, Instagram, or Reddit where members can share photos, tips, and experiences.

2. Moderate Content: Actively moderate the group to ensure a positive and respectful environment. Encourage members to post regularly and engage with each other's content.

3. Events and Challenges: Organize online events, such as photo challenges, Q&A sessions, and live cloud spotting sessions, to keep the community active and engaged.

Step 2: Cloud Spotting Websites

1. Join Existing Platforms: Participate in established cloud spotting and meteorology websites and forums, such as Cloud Appreciation Society, Weather Underground, and Met Office.

2. Contribute Content: Share your cloud observations, photos, and articles on these platforms to reach a wider audience and contribute to the broader cloud spotting community.

3. Interactive Features: Utilize interactive features like discussion boards, comment sections, and photo galleries to engage with other members.

Step 3: Online Courses and Webinars

1. Offer Courses: Develop and offer online courses or webinars on cloud spotting, cloud identification, and weather forecasting to educate and attract new members.

2. Collaborate with Experts: Partner with meteorologists and experienced cloud spotters to provide high-quality content for your courses and webinars.

3. Interactive Learning: Include interactive elements, such as quizzes, live Q&A sessions, and discussion forums, to enhance the learning experience.

15.3.3 Long-term Engagement

Step 1: Continuous Learning Opportunities

1. Resource Library: Create a digital library of books, articles, videos, and tutorials on cloud spotting and meteorology. Make these resources easily accessible to members.

2. Advanced Workshops: Organize workshops on advanced topics, such as climate science, advanced cloud photography techniques, and atmospheric dynamics.

3. Guest Speakers: Regularly invite guest speakers to share their expertise and insights with the community.

Step 2: Mentorship Programs

1. Pairing Mentors and Mentees: Establish a mentorship program where experienced cloud spotters can guide and support newcomers.

2. Regular Check-ins: Facilitate regular check-ins between mentors and mentees to discuss progress, answer questions, and provide feedback.

3. Mentorship Events: Host events specifically for mentors and mentees, such as group outings and skill-building workshops, to strengthen the mentor-mentee relationships.

Step 3: Community Projects

1. Citizen Science Initiatives: Participate in or initiate citizen science projects that involve cloud observation and data collection. Collaborate with researchers and organizations to contribute valuable data.

2. Local Exhibitions: Organize local exhibitions showcasing members' cloud photographs and projects. This can raise awareness and attract new members.

3. Seasonal Campaigns: Launch seasonal campaigns focusing on different cloud types or weather phenomena, encouraging members to participate and share their findings.

Step 4: Recognition and Rewards

1. Member Spotlights: Feature a different member each month in your newsletter or social media channels. Highlight their contributions, cloud photos, and personal stories.

2. Achievement Awards: Recognize members' achievements with awards or certificates. Categories can include most active member, best cloud photo, and most educational contribution.

3. Participation Incentives: Offer small incentives, such as cloud-themed merchandise or free event tickets, to encourage participation in activities and events.

By following these detailed steps, you can build a thriving community of cloud enthusiasts. Effective networking, leveraging online platforms, and ensuring long-term engagement will help sustain and grow your community, fostering a shared passion for cloud spotting and meteorology.

Chapter 15 Review

Chapter 15 of the HowExpert Guide to Cloud Spotting focuses on fostering a love for cloud spotting in future generations. It provides practical guidance on teaching cloud spotting to kids, creating and organizing cloud spotting clubs and events, and building a strong community of cloud enthusiasts. By following these steps, you can inspire and engage others in the fascinating world of cloud spotting.

15.1 Teaching Cloud Spotting to Kids

This section offers strategies to engage children in cloud spotting through educational activities, fun projects, and methods to sustain their interest.

- Educational Activities: Introduce cloud identification games, cloud watching sessions, and weather journals to make learning about clouds interactive and fun.

- Fun Projects: Engage kids with creative projects such as cloud art, making a cloud in a jar, and cloud photography to deepen their understanding and enjoyment of clouds.

- Building Interest: Use storytelling, field trips, and guest speakers to spark curiosity and build a sustained interest in cloud spotting.

15.2 Creating Cloud Spotting Clubs and Events

Forming a cloud spotting club and organizing events can create a vibrant community of enthusiasts.

- Starting a Club: Steps include defining the club's purpose, recruiting members, establishing a leadership team, and creating guidelines and policies.

- Organizing Events: Plan regular cloud spotting walks, host workshops and lectures, organize photo contests and exhibitions, and arrange special field trips and seasonal celebrations.

- Engaging Members: Use interactive activities, maintain effective communication channels, recognize member contributions, and provide continuous learning opportunities to keep members actively involved.

15.3 Building a Community of Cloud Enthusiasts

Building a strong community enhances the cloud spotting experience and promotes shared learning.

- Networking Tips: Organize local meetups, collaborate with other groups, and attend conferences and festivals to expand your network.

- Online Platforms: Utilize social media groups, participate in cloud spotting websites and forums, and offer or engage in online courses and webinars to connect with enthusiasts globally.

- Long-term Engagement: Provide continuous learning opportunities, establish mentorship programs, initiate community projects, and recognize and reward member achievements to sustain and grow the community.

By implementing the comprehensive strategies provided in Chapter 15, you can successfully inspire future cloud spotters, create engaging clubs and events, and build a thriving community of cloud enthusiasts. These efforts will ensure the joy and knowledge of cloud spotting are shared and sustained for generations to come.

Conclusion

Chapter 16: Embracing the Sky

As we conclude the **HowExpert Guide to Cloud Spotting**, it's essential to reflect on the journey we've taken, the beauty we've observed, and the importance of clouds in our lives. This chapter encourages readers to continue their cloud spotting adventures, maintain a lifelong interest, and inspire others to appreciate the wonders of the sky.

16.1 Reflecting on the Beauty and Importance of Clouds

Cloud spotting is more than just a hobby; it's a journey that offers endless beauty and profound insights into the natural world. This section delves into the personal reflections of cloud spotters, the joy that comes from observing clouds, and the importance of inspiring others to join in this rewarding activity.

16.1.1 Personal Reflections

Step 1: Reflect on Your Journey

1. Take Time to Look Back: Set aside some quiet time to reflect on your cloud spotting journey. Think about how you started and the milestones you've achieved along the way.

2. Memorable Moments: Identify specific moments that were particularly impactful. These could be times when you spotted a rare cloud formation, experienced a beautiful sunset, or felt a deep connection to nature.

3. Growth and Learning: Consider how your understanding of clouds and weather patterns has grown. Reflect on the skills you've developed and the knowledge you've gained.

Step 2: Document Your Reflections

1. Journaling: Write down your reflections in a journal. Include descriptions of memorable sightings, personal thoughts, and how cloud spotting has influenced your life.

2. Photography: Create a photo album or digital gallery of your favorite cloud photos. Add captions or notes to each image to document why it's special to you.

3. Storytelling: Share your journey with others through blogs, social media posts, or storytelling sessions. Your experiences can inspire and resonate with fellow cloud enthusiasts.

Step 3: Appreciate the Impact

1. Emotional Connection: Acknowledge the emotional connection you've developed with the sky. Cloud spotting can bring a sense of peace, joy, and wonder that enriches your life.

2. Mindfulness: Recognize how cloud spotting has helped you practice mindfulness. Observing clouds requires you to be present in the moment, fostering a deeper appreciation for the natural world.

16.1.2 The Joy of Observation

Step 1: Embrace the Moment

1. Mindful Observation: Practice mindful observation by focusing entirely on the clouds. Pay attention to their shapes, movements, and colors without any distractions.

2. Savor the Experience: Take time to savor the experience of cloud spotting. Whether you're alone or with others, let yourself fully enjoy the beauty of the sky.

Step 2: Create a Routine

1. Daily or Weekly Practice: Incorporate cloud spotting into your daily or weekly routine. Set aside specific times for this activity, such as early morning walks or evening relaxation periods.

2. Observation Spots: Identify favorite observation spots where you can regularly watch the clouds. This could be a local park, your backyard, or a quiet place with a clear view of the sky.

Step 3: Share the Joy

1. Family and Friends: Invite family and friends to join you in cloud spotting. Sharing this activity can deepen relationships and create shared memories.

2. Cloud Watching Events: Organize cloud watching events in your community. Gather people to observe clouds together, fostering a sense of community and shared joy.

16.1.3 Inspiring Others

Step 1: Share Your Passion

1. Talk About It: Don't hesitate to talk about your passion for cloud spotting with others. Share why you love it and what it means to you.

2. Showcase Your Work: Display your cloud photos, drawings, or journal entries at home, in the office, or on social media. Let others see the beauty you've captured.

Step 2: Educational Outreach

1. Workshops and Classes: Offer to lead workshops or classes on cloud spotting at local schools, community centers, or nature clubs. Teaching others can be a rewarding way to share your knowledge.

2. Public Presentations: Give presentations about clouds and cloud spotting at local events or online platforms. Use visuals and stories to engage and inspire your audience.

Step 3: Community Involvement

1. Start a Club: Start a cloud spotting club in your area. Provide a space for enthusiasts to gather, share experiences, and learn from each other.

2. Volunteer: Volunteer with organizations focused on environmental education and awareness. Use your cloud spotting skills to contribute to citizen science projects or community events.

Step 4: Use Social Media and Online Platforms

1. Social Media Engagement: Use social media to share your cloud spotting experiences. Post photos, interesting facts, and updates about your sightings.

2. Blogs and Vlogs: Create a blog or vlog to document your cloud spotting journey. Share tips, tutorials, and personal stories to connect with a broader audience.

By reflecting on the beauty and importance of clouds, embracing the joy of observation, and inspiring others, you can deepen your connection to this wonderful hobby and help others discover the magic of cloud spotting. This approach not only enriches your own experience but also contributes to a larger community of cloud enthusiasts who share a passion for the sky.

16.2 Encouraging Lifelong Cloud Spotting

To ensure that cloud spotting remains a fulfilling and enriching hobby throughout your life, it's important to maintain your interest, engage in continuous learning, and share your passion with others. This section provides practical steps to help you stay motivated, expand your knowledge, and inspire those around you.

16.2.1 Maintaining Interest

Step 1: Diversify Your Observations

1. Explore Different Locations: Visit various locations to observe clouds. Each environment—mountains, beaches, urban areas—offers unique cloud formations and weather patterns.

2. Observe in Different Seasons: Pay attention to how clouds change with the seasons. Each season brings distinct types of clouds and atmospheric conditions.

Step 2: Set Challenges and Goals

1. Cloud Identification Goals: Challenge yourself to identify a certain number of cloud types within a month or a season. Use a cloud identification chart to track your progress.

2. Photography Projects: Start photography projects focused on capturing different cloud types, weather phenomena, or specific cloud features. Create themed albums or online galleries.

Step 3: Join Cloud Spotting Communities

1. Local Clubs: Join or form local cloud spotting clubs to share experiences and learn from other enthusiasts.

2. Online Groups: Participate in online cloud spotting groups and forums. Engaging with a community can provide new perspectives and keep you motivated.

Step 4: Document Your Observations

1. Weather Journal: Keep a detailed weather journal where you record daily observations, photographs, and notes about the clouds you spot.

2. Blog or Social Media: Share your observations and experiences through a blog or social media platforms. This can create a sense of accountability and encourage regular engagement.

16.2.2 Continuous Learning

Step 1: Study Meteorology and Cloud Science

1. Read Books and Articles: Invest in books and read articles about meteorology, cloud formation, and atmospheric science to deepen your understanding.

2. Online Courses: Enroll in online courses or attend webinars that focus on cloud science and meteorology. Many universities and educational platforms offer free or affordable options.

Step 2: Attend Workshops and Conferences

1. Local Workshops: Participate in workshops hosted by meteorological societies, environmental groups, or educational institutions.

2. National and International Conferences: Attend conferences and symposiums dedicated to weather and climate science to stay updated on the latest research and developments.

Step 3: Experiment with Advanced Techniques

1. Time-Lapse Photography: Learn and experiment with time-lapse photography to capture the dynamic movement of clouds over time.

2. Weather Instruments: Use advanced weather instruments like barometers, anemometers, and hygrometers to collect data and correlate it with your cloud observations.

Step 4: Collaborate with Experts

1. Join Citizen Science Projects: Participate in citizen science projects that involve cloud observation and data collection. Collaborating with scientists can provide valuable insights and contribute to research.

2. Network with Meteorologists: Establish connections with meteorologists and cloud researchers. Attend their talks, follow their work, and seek opportunities to contribute to their studies.

16.2.3 Sharing Your Passion

Step 1: Educate and Inspire Others

1. Teach Classes: Offer to teach cloud spotting classes or workshops at local schools, community centers, or through online platforms.

2. Public Speaking: Give presentations on cloud spotting and meteorology at local events, libraries, or nature clubs.

Step 2: Create Content

1. Write Articles and Books: Share your knowledge and experiences by writing articles for magazines, blogs, or even publishing a book about cloud spotting.

2. Produce Videos: Create instructional and documentary-style videos about cloud spotting. Share them on platforms like YouTube to reach a wider audience.

Step 3: Organize Community Events

1. Cloud Spotting Walks: Organize regular cloud spotting walks in your community. These can be casual gatherings where participants share their knowledge and observations.

2. **Cloud Exhibitions:** Set up exhibitions of cloud photography and art in local galleries, libraries, or community centers to showcase the beauty of clouds and inspire others.

Step 4: Foster Online Engagement

1. **Social Media Campaigns:** Use social media to run campaigns that promote cloud spotting. Encourage followers to share their own cloud photos and stories.

2. **Interactive Platforms:** Develop or participate in interactive online platforms where enthusiasts can upload their cloud observations, participate in challenges, and engage in discussions.

By maintaining your interest, committing to continuous learning, and sharing your passion with others, you can ensure that cloud spotting remains a rewarding and lifelong pursuit. These strategies not only enrich your own experience but also help to cultivate a broader appreciation for the wonders of the sky.

16.3 Final Thoughts and Inspiration

As we conclude the HowExpert Guide to Cloud Spotting, it's important to leave readers with a sense of inspiration and motivation. This section provides inspirational quotes, encouraging words, and a look at the future of cloud spotting. By reflecting on these final thoughts, readers can carry forward their passion for cloud spotting and continue to explore the beauty and mysteries of the sky.

16.3.1 Inspirational Quotes

Step 1: Curate Meaningful Quotes

1. **Nature and Observation:** Select quotes that emphasize the beauty of nature and the joy of observation. For example, John Muir's "In every walk with nature, one receives far more than he seeks."

2. Clouds and Weather: Include quotes specifically about clouds and weather. For instance, Ralph Waldo Emerson's "The sky is the daily bread of the eyes."

3. Inspiration and Wonder: Choose quotes that inspire wonder and curiosity. Carl Sagan's "Somewhere, something incredible is waiting to be known" can resonate deeply with cloud spotters.

Step 2: Personal Reflections on Quotes

1. Contextualize: Provide a brief context for each quote, explaining why it is meaningful and how it relates to cloud spotting.

2. Personal Connection: Share your personal connection to the quote. Explain how it has inspired you in your cloud spotting journey.

Step 3: Visual Presentation

1. Design Elements: Use aesthetically pleasing design elements to present the quotes. Consider using cloud-themed backgrounds and elegant typography.

2. Photo Integration: Integrate your favorite cloud photos with the quotes to create a visually engaging experience.

16.3.2 Encouraging Words

Step 1: Reflect on the Journey

1. Acknowledge Efforts: Recognize the effort and dedication readers have put into learning about cloud spotting. Acknowledge their progress and achievements.

2. Celebrate Discoveries: Celebrate the discoveries and memorable moments readers have experienced throughout their cloud spotting journey.

Step 2: Emphasize the Benefits

1. Mindfulness and Peace: Highlight the mindfulness and peace that come from observing clouds. Encourage readers to continue finding solace and joy in this practice.

2. Connection to Nature: Stress the importance of maintaining a connection to nature through cloud spotting. Explain how this connection can enrich their lives and foster a deeper appreciation for the environment.

Step 3: Offer Practical Advice

1. Stay Curious: Encourage readers to stay curious and open-minded. Remind them that there is always more to learn and discover in the world of clouds.

2. Persevere: Encourage perseverance, especially during challenging weather conditions or periods of less noticeable cloud activity. Remind them that patience and persistence often lead to the most rewarding observations.

3. Share with Others: Motivate readers to share their passion for cloud spotting with others. Explain how teaching and inspiring others can amplify their own enjoyment and impact.

16.3.3 Future of Cloud Spotting

Step 1: Technological Advancements

1. Emerging Technologies: Discuss how emerging technologies like drones, high-resolution satellite imagery, and advanced meteorological instruments will enhance cloud observation and research.

2. Citizen Science: Highlight the increasing role of citizen science in cloud research. Explain how everyday cloud spotters can contribute valuable data to scientific studies.

Step 2: Global Collaboration

1. International Projects: Emphasize the importance of global collaboration in understanding and preserving the Earth's climate. Mention international projects and organizations dedicated to cloud and climate research.

2. Cross-Cultural Engagement: Encourage readers to connect with cloud spotters from different cultures and regions. Sharing diverse perspectives can enrich their understanding and appreciation of clouds.

Step 3: Educational and Outreach Initiatives

1. Expanding Outreach: Predict the growth of educational and outreach initiatives aimed at promoting cloud spotting. Schools, community centers, and online platforms will increasingly incorporate cloud education into their programs.

2. Future Generations: Stress the importance of inspiring future generations. Explain how fostering a love for cloud spotting in young people can lead to a more environmentally conscious and scientifically literate society.

Step 4: Personal Commitment

1. Lifelong Passion: Encourage readers to commit to cloud spotting as a lifelong passion. Explain how continuously engaging with this hobby can provide ongoing fulfillment and discovery.

2. Environmental Stewardship: Highlight the role of cloud spotters as stewards of the environment. Emphasize their responsibility to protect and cherish the natural world.

By reflecting on these final thoughts and inspirations, readers can carry forward their passion for cloud spotting with renewed enthusiasm and a deeper sense of purpose. The sky offers endless opportunities for observation, learning, and wonder, and by embracing this journey, readers can enrich their lives and inspire those around them.

Chapter 16 Review

Chapter 16 of the HowExpert Guide to Cloud Spotting encourages readers to reflect on the beauty and importance of clouds, maintain a lifelong interest in cloud spotting, and find inspiration in the journey. This final chapter aims to leave readers with a sense of fulfillment and motivation to continue exploring the skies.

16.1 Reflecting on the Beauty and Importance of Clouds

This section invites readers to contemplate their cloud spotting experiences and the emotional and intellectual connections they've formed with the sky.

- Personal Reflections: Readers are guided to reflect on their journey, memorable moments, and the growth they've experienced through cloud spotting. They are encouraged to document their thoughts and share their stories.

- The Joy of Observation: Emphasizing mindfulness and the joy of being present, this part encourages incorporating cloud spotting into daily routines and sharing the experience with loved ones.

- Inspiring Others: Readers are motivated to inspire others by sharing their passion, participating in educational outreach, and creating content that showcases the beauty of clouds.

16.2 Encouraging Lifelong Cloud Spotting

To ensure that cloud spotting remains a lifelong passion, this section provides practical advice on maintaining interest, continuous learning, and sharing the hobby.

- Maintaining Interest: Suggestions include exploring different locations, observing clouds in various seasons, setting challenges, and joining cloud spotting communities.

- Continuous Learning: Readers are encouraged to deepen their knowledge through books, courses, workshops, and advanced techniques like time-lapse photography and using weather instruments.

- Sharing Your Passion: This part highlights the importance of educating and inspiring others, creating content, and organizing community events to foster a broader appreciation for cloud spotting.

16.3 Final Thoughts and Inspiration

In the final section, readers are provided with inspirational quotes, encouraging words, and a glimpse into the future of cloud spotting.

- Inspirational Quotes: A curated selection of quotes about nature, observation, and clouds aims to inspire and motivate readers.

- Encouraging Words: Reflecting on the journey, celebrating achievements, and offering practical advice to stay curious, persevere, and share the passion.

- Future of Cloud Spotting: Discussing technological advancements, global collaboration, and the importance of educational and outreach initiatives to ensure that cloud spotting continues to thrive.

By exploring these reflections, maintaining a passion for cloud spotting, and finding inspiration in the journey, readers can enrich their lives and those of others. This chapter encapsulates the essence of cloud spotting, encouraging a lifelong commitment to observing and appreciating the ever-changing skies.

Appendices

Appendix A: Glossary of Cloud Terms

Understanding the terminology related to cloud spotting is crucial for both beginners and experienced enthusiasts. This appendix provides a comprehensive list of cloud-related terms, offering clear and detailed definitions to enhance your knowledge and observational skills. The glossary is designed to serve as an easy reference for both novices and experts.

A1 Comprehensive List of Cloud-Related Terms

Adiabatic Cooling: The process by which a parcel of air cools as it rises and expands in the atmosphere without exchanging heat with its surroundings.

Albedo: The measure of the reflectivity of a surface or object, such as a cloud, which reflects sunlight.

Altocumulus: Mid-level clouds characterized by white or gray patches with a grainy, rolled, or puffy appearance, often arranged in layers or waves.

Altostratus: Mid-level, gray or blue-gray clouds that cover the sky in a uniform layer, often preceding a storm.

Anvil Cloud: The flat, spreading top of a cumulonimbus cloud that has reached the tropopause, often indicating the mature stage of a thunderstorm.

Cirrostratus: Thin, high-altitude clouds that cover the sky like a veil and can create halos around the sun or moon.

Cirrus: High-altitude clouds made of ice crystals, appearing thin and wispy, often indicating fair weather but can also signify an approaching warm front.

Cumulonimbus: Towering, dense clouds associated with thunderstorms, heavy rain, lightning, hail, and tornadoes.

Cumulus: Fluffy, white clouds with a flat base and a dome-shaped top, typically associated with fair weather.

Fog: A cloud that forms at ground level, reducing visibility to less than 1 kilometer.

Front: The boundary between two different air masses, often leading to cloud formation and precipitation.

Lenticular Cloud: Lens-shaped clouds that form over mountain ranges due to orographic lift, resembling UFOs.

Mammatus: Pouch-like cloud structures that hang from the underside of a cloud, usually a cumulonimbus, indicating severe weather.

Nimbostratus: Thick, dark clouds that cover the sky and bring continuous, steady precipitation.

Orographic Lift: The lifting of air caused by its movement over a mountain or other terrain, leading to cloud formation.

Stratocumulus: Low, lumpy clouds covering the sky in patches with breaks of clear sky, usually bringing dry weather.

Stratus: Low, gray clouds that cover the sky like a blanket, often bringing overcast conditions and light precipitation.

Virga: Streaks of precipitation that fall from a cloud but evaporate before reaching the ground.

Weather Front: A boundary separating two masses of air of different densities, one warmer and often more moist than the other.

A2 Easy Reference for Beginners and Experts Alike

This glossary is structured to provide a quick and easy reference for both beginners and experts, ensuring that anyone can find the information they need to enhance their cloud spotting experience.

Step 1: User-Friendly Format

1. Alphabetical Order: The terms are organized alphabetically to make it easy to find specific definitions quickly.

2. Clear Definitions: Each term is defined in clear, concise language to ensure understanding, regardless of the reader's level of expertise.

Step 2: Visual Aids

1. Illustrations and Photos: Include illustrations or photographs next to the definitions to provide visual examples of the cloud types and phenomena described.

2. Diagrams: Use diagrams to explain more complex concepts, such as the formation of different types of clouds and weather fronts.

Step 3: Practical Applications

1. Contextual Examples: Provide examples of how each term is used in the context of cloud spotting. For instance, explain how recognizing a cumulonimbus cloud can indicate approaching thunderstorms.

2. Field Tips: Offer practical tips for identifying each type of cloud and weather phenomenon in the field.

Step 4: Cross-References

1. Related Terms: Include cross-references to related terms to help readers build a broader understanding of interconnected concepts. For example, link "Orographic Lift" to "Lenticular Cloud" and "Stratocumulus."

2. Further Reading: Suggest additional resources for readers who want to delve deeper into specific topics, such as books, articles, and online courses.

By providing a user-friendly and comprehensive glossary, this appendix ensures that readers have a valuable resource at their fingertips, enhancing their ability to identify and understand various cloud formations and meteorological phenomena. This structured and detailed approach makes the glossary an essential tool for both novice cloud spotters and seasoned experts.

Appendix B: Resources for Cloud Spotters

Having access to the right resources can greatly enhance your cloud spotting experience. This appendix lists recommended books, websites, apps, organizations, and further reading opportunities that are essential for both beginners and seasoned cloud enthusiasts.

B1 Recommended Books, Websites, and Apps

Books

1. "The Cloudspotter's Guide" by Gavin Pretor-Pinney: This book provides a comprehensive and entertaining introduction to the world of cloud spotting, filled with fascinating facts and beautiful illustrations.

2. "The Weather Book: An Easy-to-Understand Guide to the USA's Weather" by Jack Williams: A practical guide that explains various weather phenomena, including cloud formations, in an accessible way.

3. "Clouds: Learning to Observe the Atmosphere" by Eric M. Wilbur: A detailed guide that helps readers understand cloud dynamics and the atmosphere's behavior.

Websites

1. Cloud Appreciation Society: www.cloudappreciationsociety.org

   - A global community dedicated to the love of clouds, offering resources, forums, and a gallery of cloud photos.

2. Weather Underground: www.wunderground.com

   - Provides real-time weather information, including radar images and satellite views, which are useful for cloud spotting.

3. National Weather Service: www.weather.gov

- The official US government weather website, offering comprehensive weather forecasts and educational resources on meteorology.

Apps

1. CloudSpotter: An app by the Cloud Appreciation Society that helps users identify and learn about different cloud types.

2. MyRadar: A weather app providing real-time radar and weather forecasts, useful for planning cloud spotting activities.

3. Weather Underground: Offers detailed weather reports and radar images, helping cloud spotters track weather conditions accurately.

B2 Cloud Spotting Organizations and Groups

Cloud Appreciation Society

- Overview: An international organization dedicated to the love of clouds.

- Benefits: Membership provides access to a wealth of resources, including educational materials, forums, and events. Members can share their cloud photos and connect with other enthusiasts worldwide.

- Website: www.cloudappreciationsociety.org

American Meteorological Society (AMS)

- Overview: A professional organization promoting the development and dissemination of information on atmospheric and related sciences.

- Benefits: Offers resources for both professionals and amateurs, including publications, conferences, and educational programs.

- Website: www.ametsoc.org

Royal Meteorological Society (RMetS)

- Overview: A UK-based society supporting weather and climate enthusiasts and professionals through events, publications, and networking opportunities.

- Benefits: Provides access to a range of educational resources, networking events, and the latest research in meteorology.

- Website: www.rmets.org

Local Meteorological Societies and Clubs

- Overview: Many regions have local meteorological societies or cloud spotting clubs that offer resources and community for enthusiasts.

- Benefits: Local groups often organize regular meetings, workshops, and field trips, providing a hands-on approach to learning about clouds and weather.

B3 Further Reading and Learning Opportunities

Online Courses and Webinars

1. Coursera and edX: Platforms that offer online courses on meteorology, climate science, and environmental studies. Look for courses offered by universities and educational institutions.

2. The Royal Meteorological Society Webinars: Regular webinars on various topics related to weather and climate, available to members and sometimes the public.

Publications and Journals

1. "Weatherwise" Magazine: A bimonthly magazine that explores various weather phenomena, including clouds, with articles written by experts in the field.

2. "Bulletin of the American Meteorological Society" (BAMS): A scientific journal that publishes research and reviews on meteorology and related disciplines.

Workshops and Conferences

1. AMS Annual Meeting: The American Meteorological Society's annual meeting, offering sessions on various meteorological topics, including cloud research.

2. Royal Meteorological Society Events: Regular events and conferences on weather and climate topics, providing networking and learning opportunities.

Citizen Science Projects

1. GLOBE Observer: A NASA-sponsored project where volunteers can submit cloud observations to help with scientific research.

2. Zooniverse: A platform for various citizen science projects, some of which involve weather and cloud observation.

By utilizing these recommended resources, cloud spotters can deepen their understanding, enhance their observational skills, and connect with a broader community of enthusiasts. Whether through books, websites, apps, organizations, or further learning opportunities, these tools provide valuable support and enrichment for anyone passionate about cloud spotting.

Appendix C: Cloud Spotting Checklist

To make the most of your cloud spotting adventures, it's essential to be well-prepared. This appendix provides a detailed checklist of the essential items for your cloud spotting kit and key points for successful cloud identification.

C1 Essential Items for Your Cloud Spotting Kit

1. Binoculars

- Purpose: To get a closer view of distant clouds and observe details not visible to the naked eye.

- Features to Look For: Opt for binoculars with a magnification of 8x to 10x and a wide field of view.

2. Camera

- Purpose: To capture images of cloud formations for later study and enjoyment.

- Features to Look For: A camera with good zoom capabilities and high resolution. A smartphone with a good camera can also be effective.

3. Notebook and Pen

- Purpose: To record observations, including cloud types, weather conditions, and location.

- Features to Look For: A weatherproof notebook is ideal for outdoor use.

Cloud Identification Guide

- Purpose: To help identify different types of clouds and understand their characteristics.

- Features to Look For: A portable guidebook or an app on your smartphone.

Weather Apps

- Purpose: To get real-time weather updates and forecasts.

- Recommended Apps: MyRadar, Weather Underground, CloudSpotter.

Sunscreen and Hat

- Purpose: To protect yourself from sun exposure while spending extended periods outdoors.

- Features to Look For: Broad-spectrum sunscreen and a wide-brimmed hat.

7. Comfortable Clothing and Footwear

- Purpose: To ensure comfort during cloud spotting excursions.

- Features to Look For: Weather-appropriate, breathable clothing and sturdy, comfortable shoes.

8. Water and Snacks

- Purpose: To stay hydrated and energized during long observation sessions.

- Features to Look For: Reusable water bottle and non-perishable, easy-to-carry snacks.

9. Field Guide for Weather Phenomena

- Purpose: To understand other weather phenomena that may accompany different cloud types.

- Features to Look For: Comprehensive and portable guidebook.

10. Portable Chair or Blanket

- Purpose: To provide a comfortable place to sit while observing clouds.

- Features to Look For: Lightweight and easy to carry.

C2 Key Points for Successful Cloud Identification

1. Learn the Basic Cloud Types

- High-Level Clouds: Cirrus, Cirrostratus, Cirrocumulus.

- Mid-Level Clouds: Altostratus, Altocumulus.

- Low-Level Clouds: Stratus, Stratocumulus, Nimbostratus.

- Clouds with Vertical Development: Cumulus, Cumulonimbus.

2. Observe Cloud Shape and Structure

- Shape: Identify whether the cloud is flat, fluffy, wispy, or towering.

- Edges: Look at the edges of the cloud—are they well-defined or diffuse?

- Thickness: Note the thickness and density of the cloud.

3. Note the Cloud's Altitude

- High-Level Clouds: Usually above 20,000 feet.

- Mid-Level Clouds: Between 6,500 and 20,000 feet.

- Low-Level Clouds: Below 6,500 feet.

4. Pay Attention to Color and Lighting

- Color: Observe the color of the cloud—white, gray, or dark.

- Lighting: Note how the cloud interacts with sunlight. High clouds may appear bright white, while low clouds can be darker.

5. Look for Weather Patterns

- Current Weather: Identify how the current weather correlates with the cloud type (e.g., cumulus clouds often indicate fair weather).

- Forecasting: Use clouds to predict upcoming weather. For example, nimbostratus clouds often bring continuous rain.

6. Observe Movement and Change

- Movement: Note the speed and direction of cloud movement.

- Development: Observe if the cloud is growing, dissipating, or changing shape.

7. Use a Cloud Identification Guide

- Guidance: Refer to your guidebook or app to cross-check your observations and confirm the cloud type.

8. Take Detailed Notes and Photos

- Documentation: Record your observations in detail, including date, time, location, and weather conditions.

- Photos: Capture images from different angles and at various times to document changes.

9. Compare with Known Examples

- References: Compare your observations with known examples in your guidebook or app to improve accuracy.

10. Join Cloud Spotting Communities

- Engagement: Share your observations with online or local cloud spotting communities for feedback and learning.

By following this comprehensive checklist, you can ensure that you are well-prepared for your cloud spotting activities and equipped with the knowledge to accurately identify and enjoy the diverse range of cloud formations. This systematic approach will enhance your cloud spotting experience, making it more enjoyable and educational.

Appendix D: Personal Cloud Spotting Log Template

Keeping a detailed and organized log of your cloud spotting observations is essential for tracking patterns, improving your identification skills, and enhancing your overall experience. This appendix provides customizable templates for recording your observations and practical tips for maintaining an organized and detailed log.

01 Customizable Templates for Recording Your Observations

Template 1: Basic Observation Log

- Date:

- Time:

- Location:

- Weather Conditions:

- Cloud Type(s):

- Altitude Level:

- Description:

- Notes:

Template 2: Advanced Observation Log

- Date:

- Time:

- Location:

- Temperature:

- Humidity:

- Wind Speed:

- Wind Direction:

- Cloud Type(s):

- Altitude Level:

- Shape & Structure:

- Color & Lighting:

- Movement & Change:

- Notes:

Template 3: Photo Log

- Date:

- Time:

- Location:

- Cloud Type(s):

- Description:

- Photo Reference:

- Notes:

___D2 Tips for Keeping an Organized and Detailed Log___

Step 1: Set Up Your Log

1. Choose a Format: Decide whether you prefer a physical notebook, a digital document, or an app for maintaining your log.

2. Template Selection: Select the template(s) that best suit your needs. You can customize these templates further based on your preferences.

Step 2: Record Regularly

1. Consistent Entries: Make entries in your log regularly, ideally daily or after each cloud spotting session.

2. Detailed Descriptions: Provide detailed descriptions for each observation, covering all relevant aspects such as weather conditions, cloud types, and notable features.

Step 3: Use Clear and Concise Language

1. Clarity: Use clear and concise language to describe your observations. Avoid ambiguous terms and be as specific as possible.

2. Standard Terminology: Use standard meteorological terms and cloud classification names to ensure consistency and accuracy.

Step 4: Include Visual References

1. Photographs: Take clear photographs of clouds and reference them in your log. Label each photo with the date, time, and location.

2. Drawings: If you enjoy sketching, include drawings of cloud formations. This can help you remember specific details and improve your observation skills.

Step 5: Track Changes Over Time

1. Compare Observations: Regularly review your past entries to identify patterns and changes in cloud formations and weather conditions.

2. Seasonal Patterns: Pay attention to seasonal changes and how they affect cloud types and weather patterns.

Step 6: Analyze Weather Data

1. Weather Conditions: Note the temperature, humidity, wind speed, and direction for each observation. This data can provide context for your cloud observations.

2. Correlation: Look for correlations between weather conditions and cloud formations to enhance your understanding of meteorological phenomena.

Step 7: Make Notes of Unusual Observations

1. Rare Phenomena: Highlight any rare or unusual cloud formations or weather phenomena in your log.

2. Additional Information: Include any additional information or observations that might be relevant, such as interactions with wildlife or interesting atmospheric effects.

Step 8: Organize Your Log Periodically

1. Review and Update: Periodically review and update your log to ensure it remains organized and comprehensive.

2. Indexing: Consider creating an index or table of contents for your log, especially if it spans multiple volumes or files.

Step 9: Share and Collaborate

1. Community Sharing: Share your observations with cloud spotting communities or online forums to gain feedback and connect with other enthusiasts.

2. Collaborative Projects: Participate in collaborative cloud spotting projects or citizen science initiatives to contribute your data to larger research efforts.

By following these detailed steps and using the provided templates, you can maintain an organized and detailed cloud spotting log. This systematic approach will help you track your observations accurately, identify patterns, and deepen your understanding of cloud formations and weather phenomena.

About the Author

HowExpert publishes how to guides on all topics from A to Z by everyday experts. Visit HowExpert.com to learn more.

About the Publisher

Byungjoon "BJ" Min is an author, publisher, entrepreneur, and the founder of HowExpert. He started off as a once broke convenience store clerk to eventually becoming a fulltime internet marketer and finding his niche in publishing. He is the founder and publisher of HowExpert where the mission is to discover, empower, and maximize everyday people's talents to ultimately make a positive impact in the world for all topics from A to Z. Visit BJMin.com and HowExpert.com to learn more. John 14:6

Recommended Resources

- HowExpert.com – How To Guides on All Topics from A to Z by Everyday Experts.
- HowExpert.com/free – Free HowExpert Email Newsletter.
- HowExpert.com/books – HowExpert Books
- HowExpert.com/courses – HowExpert Courses
- HowExpert.com/clothing – HowExpert Clothing
- HowExpert.com/membership – HowExpert Membership Site
- HowExpert.com/affiliates – HowExpert Affiliate Program
- HowExpert.com/jobs – HowExpert Jobs
- HowExpert.com/writers – Write About Your #1 Passion/Knowledge/Expertise & Become a HowExpert Author.
- HowExpert.com/resources – Additional HowExpert Recommended Resources
- YouTube.com/HowExpert – Subscribe to HowExpert YouTube.
- Instagram.com/HowExpert – Follow HowExpert on Instagram.
- Facebook.com/HowExpert – Follow HowExpert on Facebook.
- TikTok.com/@HowExpert – Follow HowExpert on TikTok.

Made in the USA
Middletown, DE
07 December 2024

66277881R00181